Beginning MLOps with MLFlow

Deploy Models in AWS SageMaker, Google Cloud, and Microsoft Azure

Sridhar Alla
Suman Kalyan Adari

Apress®

Beginning MLOps with MLFlow

Sridhar Alla
Delran, NJ, USA

Suman Kalyan Adari
Tampa, FL, USA

ISBN-13 (pbk): 978-1-4842-6548-2
https://doi.org/10.1007/978-1-4842-6549-9

ISBN-13 (electronic): 978-1-4842-6549-9

Managing Director, Apress Media LLC: Welmoed Spahr
Acquisitions Editor: Celestin Suresh John
Development Editor: Laura Berendson
Coordinating Editor: Aditee Mirashi

Cover designed by eStudioCalamar

Cover image designed by Freepik (www.freepik.com)

Distributed to the book trade worldwide by Springer Science+Business Media New York, 1 New York Plaza, Suite 4600, New York, NY 10004-1562, USA. Phone 1-800-SPRINGER, fax (201) 348-4505, e-mail orders-ny@springer-sbm.com, or visit www.springeronline.com. Apress Media, LLC is a California LLC and the sole member (owner) is Springer Science + Business Media Finance Inc (SSBM Finance Inc). SSBM Finance Inc is a **Delaware** corporation.

For information on translations, please e-mail booktranslations@springernature.com; for reprint, paperback, or audio rights, please e-mail bookpermissions@springernature.com.

Apress titles may be purchased in bulk for academic, corporate, or promotional use. eBook versions and licenses are also available for most titles. For more information, reference our Print and eBook Bulk Sales web page at www.apress.com/bulk-sales.

Any source code or other supplementary material referenced by the author in this book is available to readers on GitHub via the book's product page, located at www.apress.com/978-1-4842-6548-2. For more detailed information, please visit www.apress.com/source-code.

Printed on acid-free paper

Table of Contents

About the Authors

 **Sridhar Alla** is the founder and CTO of Bluewhale.one, the company behind the product Sas2Py (`www.sas2py.com`), which focuses on the automatic conversion of SAS code to Python. Bluewhale also focuses on using AI to solve key problems ranging from intelligent email conversation tracking to issues impacting the retail industry and more. He has deep expertise in building AI-driven big data analytical practices on both the public cloud and in-house infrastructures. He is a published author of books and an avid presenter at numerous Strata, Hadoop World, Spark Summit, and other conferences. He also has several patents filed with the US PTO on large-scale computing and distributed systems. He has extensive hands-on experience in most of the prevalent technologies, including Spark, Flink, Hadoop, AWS, Azure, TensorFlow, and others. He lives with his wife, Rosie, and daughters, Evelyn and Madelyn, in New Jersey, United States, and in his spare time loves to spend time training, coaching, and attending meetups. He can be reached at `sid@bluewhale.one`.

ABOUT THE AUTHORS

Suman Kalyan Adari is a current Senior and undergraduate researcher at the University of Florida specializing in deep learning and its practical use in various fields such as computer vision, adversarial machine learning, natural language processing (conversational AI) , anomaly detection, and more. He was a presenter at the IEEE Dependable Systems and Networks International Conference workshop on Dependable and Secure Machine Learning held in Portland, Oregon, United States in June 2019. He is also a published author, having worked on a book focusing on the uses of deep learning in anomaly detection. He can be reached at `sadari@ufl.edu`.

About the Technical Reviewer

Pramod Singh is a Manager, Data Science at Bain & Company. He has over 11 years of rich experience in the Data Science field working with multiple product- and service-based organizations. He has been part of numerous ML and AI large scale projects. He has published three books on large scale data processing and machine learning. He is also a regular speaker at major AI conferences such as the O'Reilly AI & Strata conference.

Acknowledgments

Sridhar Alla

I would like to thank my wonderful wife, Rosie Sarkaria, and my beautiful, loving daughters, Evelyn and Madelyn, for all the love and patience during the many months I spent writing this book. I would also like to thank my parents, Ravi and Lakshmi Alla, for all the support and encouragement they continue to bestow upon me.

Suman Kalyan Adari

I would like to thank my parents, Venkata and Jyothi Adari, and my loving dog, Pinky, for supporting me throughout the entire process.
I would especially like to thank my sister, Niharika Adari, for helping me with edits and proofreading and helping me write the appendix chapter.

Introduction

This book is intended for all audiences ranging from beginners at machine learning, to advanced machine learning engineers, or even to machine learning researchers who wish to learn how to better organize their experiments.

The first two chapters cover the premise of the problem followed by the book, which is that of integrating MLOps principles into an anomaly detector model based on the credit card dataset. The third chapter covers what MLOps actually is, how it works, and why it can be useful.

The fourth chapter goes into detail about how you can implement and utilize MLFlow in your existing projects to reap the benefits of MLOps with just a few lines of code.

The fifth, sixth, and seventh chapters all go over how you can operationalize your model and deploy it on AWS, Microsoft Azure, and Google Cloud, respectively. The seventh chapter goes over how you can host a model on a virtual machine and connect to the server from an external source to make your predictions, so should any MLFlow functionality described in the book become outdated, you can always go for this approach and simply serve models on some cluster on the cloud.

The last chapter, Appendix, goes over how you can utilize Databricks, the creators of MLFlow, to organize your MLFlow experiments and deploy your models.

The goal of the book is to hopefully impart to you, the reader, knowledge of how you can use the power of MLFlow to easily integrate MLOps principles into your existing projects. Furthermore, we hope that you will become more familiar with how you can deploy your models to the cloud, allowing you to make model inferences anywhere on the planet so as long as you are able to connect to the cloud server hosting the model.

At the very least, we hope that more people do begin to adopt MLFlow and integrate it into their workflows, since even as a tool to organize your workspace, it massively improves the management of your machine learning experiments and allows you to keep track of the entire model history of a project.

Researchers may find MLFlow to be useful when conducting experiments, as it allows you to log plots on top of any custom-defined metric of your choosing. Prototyping becomes much easier, as you can now keep track of that one model which worked perfectly as a proof-of-concept and revert back to those same weights at any time while you keep tuning the hyperparameters. Hyperparameter tuning becomes much simpler and more organized, allowing you to run a complex script that searches over several different hyperparameters at once and log all of the results using MLFlow.

With all the benefits that MLFlow and the corresponding MLOps principles offer to machine learning enthusiasts of all professions, there really are no downsides to integrating it into current work environments. With that, we hope you enjoy the rest of the book!

CHAPTER 1

Getting Started: Data Analysis

In this chapter, we will go over the premise of the problem we are attempting to solve with the machine learning solution we want to operationalize. We will also begin data analysis and feature engineering of our data set.

Introduction and Premise

Welcome to *Beginning MLOps with MLFlow*! In this book, we will be taking an example problem, developing a machine learning solution to it, and operationalizing our model on AWS SageMaker, Microsoft Azure, Google Cloud, and Datarobots. The problem we will be looking at is the issue of performing anomaly detection on a credit card data set. In this chapter, we will explore this data set and show the overall structure while explaining a few techniques on analyzing this data. This data set can be found at `www.kaggle.com/mlg-ulb/creditcardfraud`.

If you are already familiar with how to analyze data and build machine learning models, feel free to grab the data set and skip ahead to 3 to jump right into MLOps.

© Sridhar Alla, Suman Kalyan Adari 2021
S. Alla and S. K. Adari, *Beginning MLOps with MLFlow*,
https://doi.org/10.1007/978-1-4842-6549-9_1

Otherwise, we will first go over the general process of how machine learning solutions are generally created. The process goes something like this:

1. **Identification of the problem:** First of all, you need to have an idea of what the problem is, what can be done about it, what has been done about it, and why it is a problem worth solving.

   Here's an example of a problem: an invasive snake species harmful to the local environment has infested a region. This species is highly venomous and looks very similar to a harmless species of snake native to this same environment. Furthermore, the invasive species is destructive to the local environment and is outcompeting the local species.

   In response, the local government has issued a statement encouraging citizens to go out and kill the venomous, invasive species on sight, but the problem is that it turns out citizens have been killing the local species as well due to how easy it is to confuse the two species.

   What can be done about this? A possible solution is to use the power of machine learning and build an application to help citizens identify the snake species. What has been done about it? Perhaps someone released an app that does a poor job at distinguishing the two species, which doesn't help remedy the current situation. Perhaps fliers have been given out, but it can be hard to identify every member of a species correctly based on just one picture.

Why is it a problem worth solving? The native species is important to the local environment. Killing the wrong species can end up exacerbating the situation and lead to the invasive species claiming the environment over the native species. And so building a computer vision-based application that can discern between the various snake species (and especially the two species relevant to the problem) could be a great way to help citizens get rid of the right snake species.

2. **Collection of data:** After you've identified the problem, you want to collect the relevant data. In the context of the snake species classification problem, you want to find images of various snake species in your region. The location depends on how big of a scale your project will operate on. Is it going to identify any snake in the world? Just snakes in Florida?

   If you can afford to do so, the more data you collect, the better the potential training outcomes will be. More training examples can introduce increased variety to your model, making it better in the long run. Deep learning models scale in performance with large volumes of data, so keep that in mind as well.

3. **Data analysis:** Once you've collected all the raw data, you want to clean it up, process it, and format it in a way that allows you to analyze the data better.

   For images, this could be something like applying an algorithm to crop out unnecessary parts of the image to focus solely on the snake. Additionally, maybe you want to center-crop the image to remove all the extra visual information in the data sample. Either way, raw image data is rarely ever in good enough condition to be used directly; it almost always requires processing to get the relevant data you want.

For unstructured data like images, formatting this data in a way good enough to analyze it could be something like creating a directory with all of the respective snake species and the relevant image data. From there, you can look at the count of images for each snake species class that you have and determine if you need to retrieve more samples for a particular species or not.

For structured data, say the credit-card data set, processing the raw data can mean something like getting rid of any entries with null values in them. Formatting them in a way so you can analyze them better can involve dimensionality-reduction techniques such as principal component analysis (PCA). Note: It turns out that most of the data in the credit card data set has actually been processed with PCA in part to preserve the privacy of the users the data has been extracted from.

As for the analysis, you can construct multiple graphs of different features to get an idea of the overall distribution and how the features look plotted against each other. This way, you can see any significant relationships between certain features that you might keep in mind when creating your training data.

There are some tools you can use in order to find out what features have the greatest influence on the label, such as **phi-k correlation**. By allowing you to see the different correlation values between the individual features and the target label, you can gain a deeper understanding of the relationships between the features in this data set. If needed, you can also drop features that aren't very influential from the data. In this step, you really want to get a solid understanding of your data so you can apply a model architecture that is most suitable for it.

4. **Feature engineering and data processing:** Now you can use the knowledge you gained from analyzing the various features and their relationships to each another to potentially construct new features from combinations of several existing ones. For example, the Titanic data set is a great example that you can apply feature engineering to. In this case, you can take information such as class, age, fare, number of siblings, number of parents, and so on to create as many features as you can think up.

    Feature engineering is really about giving your model a deeper context so it can learn the task better. You don't necessarily want to create random features for the sake of it, but something that's potentially relevant like number of female relatives, for example. (Since females were more likely to survive the sinking of the Titanic, could it be possible that if a person had more female relatives, they were less likely to survive as preference was given to their female relatives instead?)

The next step after feature engineering is data processing, which is a step involving all preparations made to process the data to be passed into the model. In the context of the snake species image data, this could involve normalizing all the values to be between 0 and 1 as well as "batching" the data into groups.

This step also usually creates several subsets of your initial data: a **training data set**, a **testing data set**, and a **validation data set**. We will go into more detail on the purpose of each of these data sets later. For now, a **training data set** contains the data you want the model to learn from, the **testing data set** contains data you want to evaluate the model's performance on, and the **validation data set** is used to either select a model or help tune a model's hyperparameters to draw out a better performance.

5. **Build the model:** Now that the data processing is done, this step is all about selecting the proper architecture and building the model. For the snake species image data, a good choice would be to use a convolutional neural network (CNN) because they work very well for any tasks involving images. From there, it is up to you to define the specific architecture of the model with respect to its layer composition.

6. **Training, evaluating, and validating:** When you're training your CNN model, you're usually passing in batches of data until the entire data makes a full pass through the model. From the results of this "forward pass," calculations are made that tell the model how to adjust the weights as they are made going backwards across the network in what's called the "backward

pass." The training process is essentially where the model learns how to perform the task and gets better at it the more examples it sees.

After the training process, either the evaluation step or the validation step can come next. As long as the testing set and validation set come from different distributions (the validation set can be derived from the training set, while the testing set can be derived from the original data), the model is technically seeing new data in the evaluation and validation processes. The model will never learn anything from the evaluation data, so you can test your model anytime.

Model evaluation is where the model's performance metrics such as accuracy, precision, recall, and so on are evaluated on a data set that it has never seen before. We will go into more detail on the evaluation step once it becomes more relevant in the next chapter, Chapter 2.

Depending on the context, the exact purpose of validation can differ, along with the question of whether or not evaluation should be performed first after training. Let's define several sample scenarios where you would use validation:

- **Selecting a model architecture:** Of several model types or architectures, you use k-fold cross-validation, for example, to quickly train and evaluate each of the models on some data partition of the validation set to get an idea of how they are performing. This way, you can get a good idea of which model is performing best, allowing you to pick a model and continue with the rest of the process.

- **Selecting the best model:** Of several trained models, you can use something like k-fold cross-validation to quickly evaluate each model on the validation data to allow you to get an idea of which ones are performing best.

- **Tuning hyperparameters:** Quickly train a model and test it with different hyperparameter setups to get an idea of which configurations work better. You can start with a broad range of hyperparameters. From there, you can use the results to narrow the range of hyperparameters until you get to a configuration where you are satisfied. Models in deep learning, for example, can have many hyperparameters, so using validation to tune those hyperparameters can work well in deep learning settings. Just beware of diminishing returns. After a certain precision with the hyperparameter setting, you're not going to see that big of a performance boost in the model.

- **Indication of high variance:** This validation data is slightly different from the other three examples. In the case of neural networks, this data is derived from a small split of the training data. After one full pass of the training data, the model evaluates on this validation data to calculate metrics such as loss and accuracy.

  If your training accuracy is high and training loss is low, but the validation accuracy is low and the validation loss is high, that's an indication that your model suffers from **high variance**. What this means is that your model has not learned to generalize what it is "learning" to new data, as the validation data in this case is comprised of data it has never seen before. In other words, your model is **overfitting**. The model just isn't recreating the kind of performance it gets on the training data on new data that it hasn't seen before.

If your model has poor training accuracy and high training loss, then your model suffers from **high bias**, meaning it isn't learning how to perform the task correctly on the training data at all.

This little validation split during the training process can give you an early indication of when overfitting is occurring.

7. **Predicting:** Once the model has been trained, evaluated, and validated, it is then ready to make predictions. In the context of the snake species detector, this step involves passing in visual images of the snake in question to get some prediction back. For example, if the model is supposed to detect the snake, draw a box around it, and label it (in an object detection task), it will do so and display the results in real time in the application.

If it just classifies the snake in the picture, the user simply sends their photo of a snake to the model (via the application) to get a species classification prediction along with perhaps a probability confidence score.

Hopefully now you have a better idea of what goes on when creating machine learning solutions.

With all that in mind, let's get started on the example, where you will use the credit card data set to build simple anomaly detection models using the data.

Credit Card Data Set

Before you perform any data analysis, you need to first collect your data. Once again, the data set can be found at the following link: `www.kaggle.com/mlg-ulb/creditcardfraud`.

Following the link, you should see something like the following in Figure 1-1.

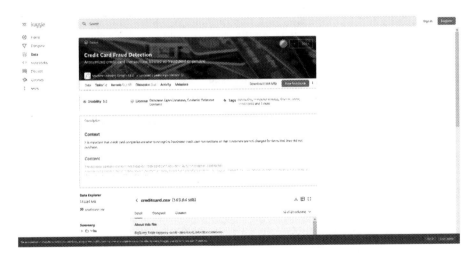

Figure 1-1. *Kaggle website page on the credit card data*

From here, you want to download the data set by clicking the Download (144 MB) button next to New Notebook. It should take you to a sign-in page if you're not already signed in, but you should be able to download the data set after that.

Once the zip file finishes downloading, simply extract it somewhere to reveal the credit card data set. Now let's open up Jupyter and explore this data set. Before you start this step, let's go over the exact packages and their versions:

- Python 3.6.5

- numpy 1.18.3

- pandas 0.24.2

- matplotlib 3.2.1

To check your package versions, you can run a command like

```
pip show package_name
```

Alternatively, you can run the following code to display the version in the notebook itself:

```
import module_name
print(module_name.__version__)
```

In this case, `module_name` is the name of the package you're importing, such as numpy.

Loading the Data Set

Let's begin! First, open a new notebook and import all of the dependencies and set global parameters for this notebook:

```
%matplotlib inline

import numpy as np
import pandas as pd
import matplotlib.pyplot as plt
from pylab import rcParams

rcParams['figure.figsize'] = 14, 8
```

Refer to Figure 1-2.

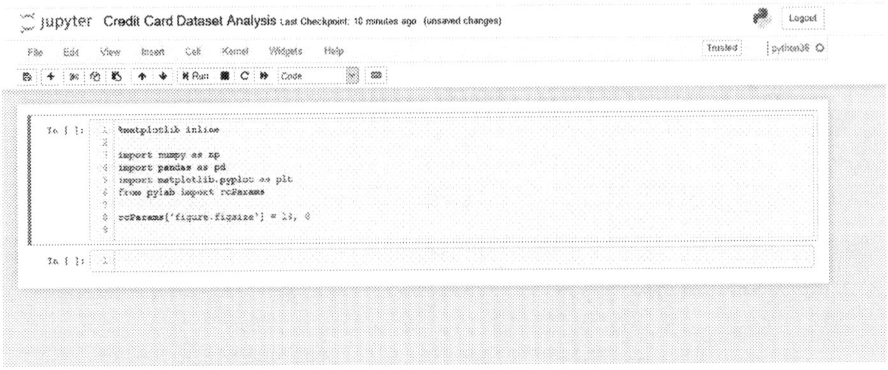

Figure 1-2. *Jupyter notebook cell with some import statements as well as a global parameter definition for the size of all matplotlib plots*

Now that you have imported the necessary libraries, you can load the data set. In this case, the data folder exists in the same directory as the notebook file and contains the creditcard.csv file. Here is the code:

```
data_path = "data/creditcard.csv"

df = pd.read_csv(data_path)
```

Refer to Figure 1-3.

Figure 1-3. *Defining the data path to the credit card data set .csv file, reading its contents, and creating a pandas data frame object*

Now that the data frame has been loaded, let's take a look at its contents:

df.head()

Refer to Figure 1-4.

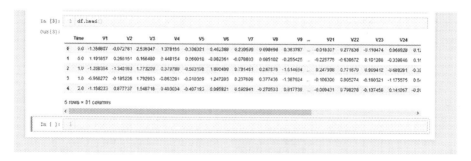

Figure 1-4. *Calling the head() function on the data frame to display the first five rows of the data frame*

If you are not familiar with the `df.head(n)` function, it essentially prints the first n rows of the data frame. If you did not pass any arguments, like in the figure above, then the function defaults to a value of five, printing the first five rows of the data frame.

Feel free to play around with that function as well as use the scroll bar to explore the rest of the features.

Now, let's look at some basic statistical values relating to the values in this data frame:

```
df.describe()
```

Refer to Figure 1-5.

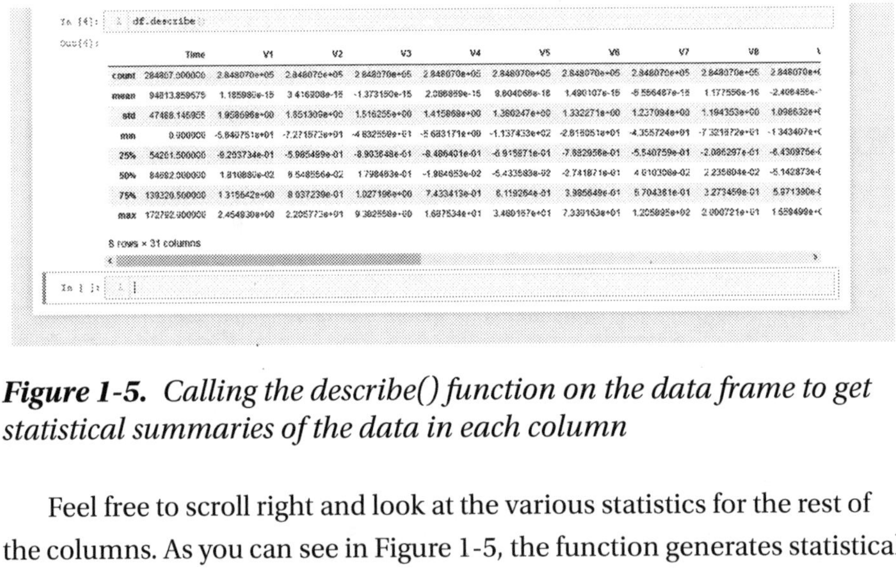

In [4]: df.describe()

Out[4]:

	Time	V1	V2	V3	V4	V5	V6	V7	V8	\
count	284807.000000	2.848070e+05	2.848070e+05	2.848070e+05	2.848070e+05	2.848070e+05	2.848070e+05	2.848070e+05	2.848070e+05	2.848070e+(
mean	94813.859575	1.165980e-15	3.416908e-15	-1.373150e-15	2.086859e-15	9.604066e-16	1.490107e-15	-5.556467e-16	1.177556e-16	-2.406455e-
std	47488.145955	1.958696e+00	1.651309e+00	1.516255e+00	1.415869e+00	1.380247e+00	1.332271e+00	1.237094e+00	1.194353e+00	1.098632e+(
min	0.000000	-5.640751e+01	-7.271573e+01	-4.832559e+01	-5.683171e+00	-1.137433e+02	-2.616051e+01	-4.355724e+01	-7.321672e+01	-1.343407e+(
25%	54201.500000	-9.203734e-01	-5.985499e-01	-8.903648e-01	-8.486401e-01	-6.915971e-01	-7.682956e-01	-5.540759e-01	-2.086297e-01	-6.430976e-(
50%	84692.000000	1.810880e-02	6.548556e-02	1.798463e-01	-1.984653e-02	-5.433583e-02	-2.741871e-01	4.010308e-02	2.235804e-02	-5.142873e-(
75%	139320.500000	1.315642e+00	8.037239e-01	1.027196e+00	7.433413e-01	6.119264e-01	3.985649e-01	5.704361e-01	3.273459e-01	5.971390e-(
max	172792.000000	2.454930e+00	2.205773e+01	9.382558e+00	1.687534e+01	3.480167e+01	7.330163e+01	1.205895e+02	2.000721e+01	1.559499e+(

8 rows × 31 columns

In [ ]:

***Figure 1-5.** Calling the describe() function on the data frame to get statistical summaries of the data in each column*

Feel free to scroll right and look at the various statistics for the rest of the columns. As you can see in Figure 1-5, the function generates statistical summaries for data in each of the columns in the data frame.

The main takeaway here is that there are a huge number of data points. In fact, you can check the shape of the data frame by simply calling

```
df.shape
```

Refer to Figure 1-6.

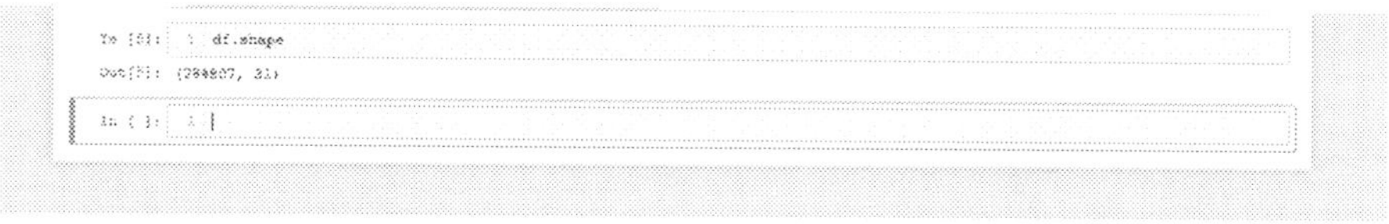

Figure 1-6. *Calling the shape() function on the data frame to get an output in the format (number_of_rows, number_of_columns)*

There are 284,807 rows and 31 columns in this data frame. That's a lot of entries! Not only that, but if you look at Figure 1-5, you'll see that the values can get really large for the column `Time`. In fact, keep scrolling right, and you'll see that values can get very large for the column `Amount` as well. Refer to Figure 1-7.

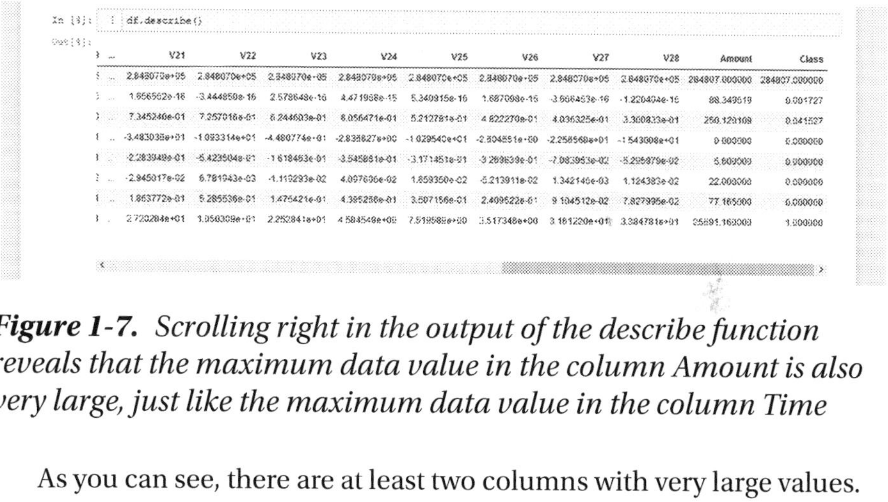

...	V21	V22	V23	V24	V25	V26	V27	V28	Amount	Class
...	2.848070e+05	2.848070e+05	2.848070e+05	2.848070e+05	2.848070e+05	2.848070e+05	2.848070e+05	2.848070e+05	284807.000000	284807.000000
...	1.656562e-16	-3.444850e-16	2.578648e-16	4.471968e-15	5.340915e-16	1.687098e-15	-3.666453e-16	-1.220404e-16	88.349619	0.001727
...	7.345240e-01	7.257016e-01	6.244603e-01	6.056471e-01	5.212781e-01	4.822270e-01	4.036325e-01	3.300833e-01	250.120109	0.041527
...	-3.483038e+01	-1.093314e+01	-4.480774e+01	-2.836627e+00	-1.029540e+01	-2.604551e+00	-2.256568e+01	-1.543008e+01	0.000000	0.000000
...	-2.283949e-01	-5.423504e-01	-1.618463e-01	-3.545861e-01	-3.171451e-01	-3.269839e-01	-7.083953e-02	-5.295979e-02	5.600000	0.000000
...	-2.945017e-02	6.781943e-03	-1.119293e-02	4.097606e-02	1.659350e-02	-5.213911e-02	1.342146e-03	1.124383e-02	22.000000	0.000000
...	1.863772e-01	5.285536e-01	1.476421e-01	4.395266e-01	3.507156e-01	2.409522e-01	9.104512e-02	7.827995e-02	77.165000	0.000000
...	2.720284e+01	1.050309e+01	2.252841e+01	4.584549e+00	7.519589e+00	3.517346e+00	3.161220e+01	3.384781e+01	25691.160000	1.000000

Figure 1-7. *Scrolling right in the output of the describe function reveals that the maximum data value in the column Amount is also very large, just like the maximum data value in the column Time*

As you can see, there are at least two columns with very large values. What this tells you is that later on, when building the various data sets for the model training process, you definitely need to scale down the data. Otherwise, such large data values can potentially mess up the training process.

Normal Data and Fraudulent Data

Since there are only two classes, normal and fraud, let's split up the data frame by class and continue with the data analysis. In the context of anomaly detection, the fraud class is also the anomaly class, hence why we chose to name the data frame representing fraudulent transaction data anomalies and interchangeably refer to this class as either fraud or anomaly.

Here is the code:

```
anomalies = df[df.Class == 1]
normal = df[df.Class == 0]
```

After that, run the following in a separate cell:

```
print(f"Anomalies: {anomalies.shape}")
print(f"Normal: {normal.shape}")
```

Refer to Figure 1-8.

Figure 1-8. *Defining data frames for fraudulent/anomalous data and for normal data and printing their shapes*

From here, you can see that the data is overwhelmingly biased towards normal data, and that anomalies only comprise a vast minority of data points in the overall data set. What this tells you is that you will have to craft the training, evaluation, and validation sets more carefully so each of these sets will have a good representation of anomaly data.

In fact, let's look at this disparity in a graphical manner just to see how large the difference is:

```
class_counts = pd.value_counts(df['Class'], sort = True)
class_counts.plot(kind = 'bar', rot=0)
plt.title("Class Distribution")
plt.xticks(range(2), ["Normal", "Anomaly"])
plt.xlabel("Label")
plt.ylabel("Counts")
```

Refer to Figure 1-9.

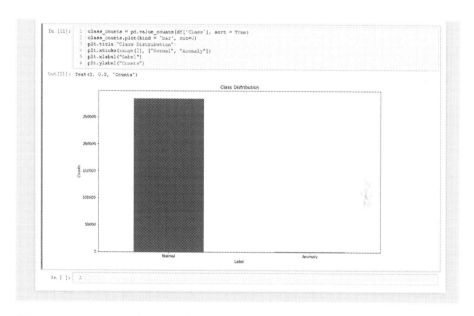

Figure 1-9. *A graph visually demonstrating the difference in counts for normal data and anomalous data*

The graph visually shows the immense difference between the number of data values of the two classes.

So now you can begin analyzing some of the characteristics of data points in each class. First of all, the columns in this data set are Time, values V1 through V28, Amount, and Class.

So, do anomalous data values comprise transactions with excessive amounts? Let's look at some statistical summary values for Amount:

```
anomalies.Amount.describe()
```

Refer to Figure 1-10 for the output.

Figure 1-10. *Output of the describe() function on the data frame for fradulent values for the column Amount*

It seems like the data is skewed right, and that anomalous transactions comprise values that are not very high. In fact, most of the transactions are less than $100, so it's not like fraudulent transactions are high-value transactions.

```
normal.Amount.describe()
```

Refer to Figure 1-11 for the output.

Figure 1-11. *Output of the describe() function on the data frame for normal values for the column Amount*

If you look at the normal data, it's even more skewed right than the anomalies. Most of the transactions are below $100, and some of the amounts can get very high to values like $25,000.

Plotting

Let's now turn to a graphical approach to help visually illustrate this better. First, you define some functions to help plot the various columns of the data to make it much easier to visualize the various relationships:

```python
def plot_histogram(df, bins, column, log_scale=False):

    bins = 100

    anomalies = df[df.Class == 1]
    normal = df[df.Class == 0]

    fig, (ax1, ax2) = plt.subplots(2, 1, sharex=True)
    fig.suptitle(f'Counts of {column} by Class')

    ax1.hist(anomalies[column], bins = bins, color="red")
    ax1.set_title('Anomaly')

    ax2.hist(normal[column], bins = bins, color="orange")
    ax2.set_title('Normal')

    plt.xlabel(f'{column}')
    plt.ylabel('Count')
    if log_scale:
        plt.yscale('log')
    plt.xlim((np.min(df[column]), np.max(df[column])))
    plt.show()
```

```python
def plot_scatter(df, x_col, y_col, sharey = False):

    anomalies = df[df.Class == 1]
    normal = df[df.Class == 0]

    fig, (ax1, ax2) = plt.subplots(2, 1, sharex=True,
    sharey=sharey)
    fig.suptitle(f'{y_col} over {x_col} by Class')

    ax1.scatter(anomalies[x_col], anomalies[y_col], color='red')
    ax1.set_title('Anomaly')

    ax2.scatter(normal[x_col], normal[y_col], color='orange')
    ax2.set_title('Normal')

    plt.xlabel(x_col)
    plt.ylabel(y_col)
    plt.show()
```

Refer to Figure 1-12 to see the code in cells.

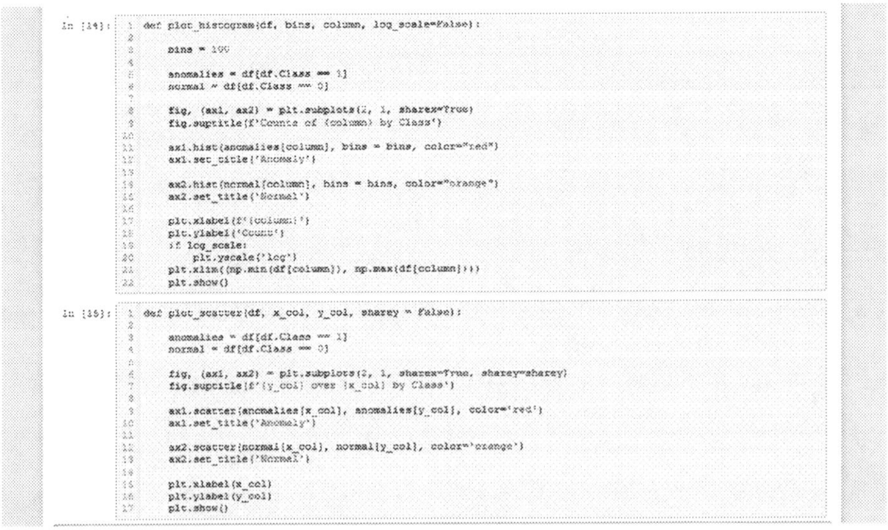

Figure 1-12. *Each of the plotter functions in their own Jupyter cells*

Now, let's start by plotting values for Amount by Class for the entire data frame:

```
plt.scatter(df.Amount, df.Class)
plt.title("Transaction Amounts by Class")
plt.ylabel("Class")
plt.yticks(range(2), ["Normal", "Anomaly"])
plt.xlabel("Transaction Amounts ($)")
plt.show()
```

Refer to Figure 1-13.

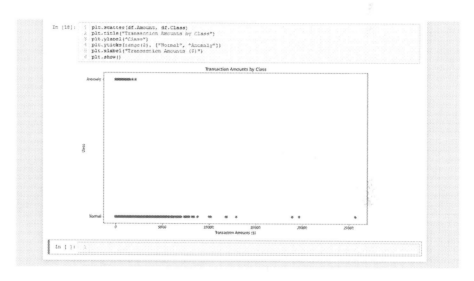

Figure 1-13. *A scatterplot of data values in the data frame encompassing all the data values. The plotted columns are Amount on the x-axis and Class on the y-axis*

It seems like there are some massive outliers in the normal data set, as suspected. However, the graph isn't very informative in telling you about

value counts, so let's use the plotting functions defined earlier to draw
graphs that provide more context:

```
bins = 100

plot_histogram(df, bins, "Amount", log_scale=True)
```

Refer to Figure 1-14.

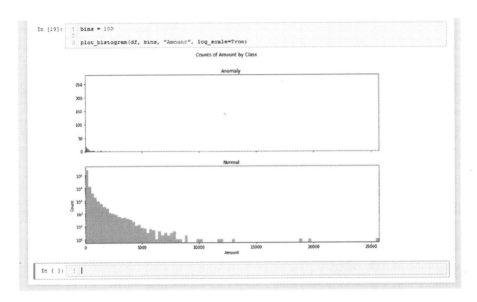

Figure 1-14. *A histogram of counts for data values organized into
intervals in the column Amount in the data frame. The number of
bins is 100, meaning the interval of each bar in the histogram is the
range of the data in the column Amount divided by the number of bins*

From this, you can definitely notice a right skew as well as the massive
outliers present in the normal data. Since you can't really see much of the
anomalies, let's create another plot:

```
plt.hist(anomalies.Amount, bins = bins, color="red")
plt.show()
```

Refer to Figure 1-15.

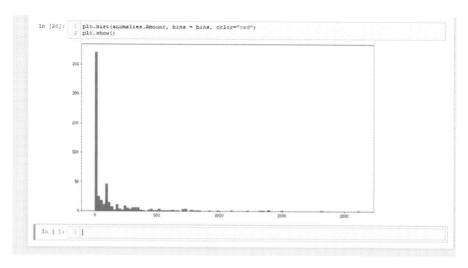

Figure 1-15. *A histogram of just the values in the anomaly data frame for the column Amount. The number of bins is also 100 here, as it will be for the rest of the examples*

The anomalies seem to be right skewed as well, but much more heavily so. This means that the majority of anomalous transactions actually have quite low transaction amounts.

Alright, so what about time? Let's plot another basic scatterplot:

```
plt.scatter(df.Time, df.Class)
plt.title("Transactions over Time by Class")
plt.ylabel("Class")
plt.yticks(range(2), ["Normal", "Anomaly"])
plt.xlabel("Time (s)")
plt.show()
```

Refer to Figure 1-16.

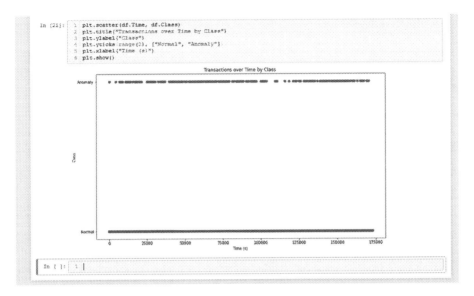

Figure 1-16. *A scatterplot for values in the data frame df with data in the column Time on the x-axis and data in the column Class in the y-axis*

This graph isn't very informative, but it does tell you that fraudulent transactions are pretty spread out over the entire timeline. Once again, let's use the plotter functions to get an idea of the counts:

```
plot_scatter(df, "Time", "Amount")
```

Refer to Figure 1-17.

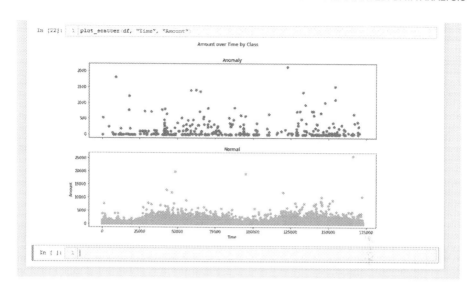

Figure 1-17. *Using the plot_scatter() function to plot data values for the columns Time on the x-axis and Amount on the y-axis in the df data frame*

You have a better context now, but it doesn't seem to tell you much. You can see that fraudulent transactions occur throughout the entire timeline and that there is no specific period of time when it seems like higher-value transactions occur. There do seem to be two main clusters, but this could also be a result of the lack of data points compared to the normal points.

Let's now look at the histogram to take into account frequencies:

```
plot_scatter(df, "Time", "Amount")
```

Refer to Figure 1-18.

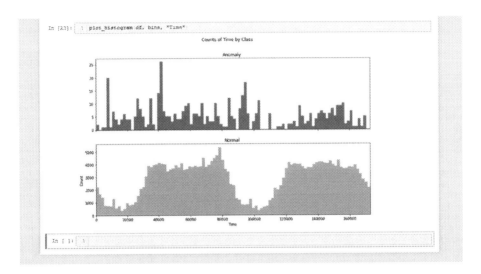

Figure 1-18. *Using the plot_histogram() function to plot data values for the column Time in the df data frame*

From this, you get a really good context of the amount of fraudulent/anomalous transactions going on over time. For the normal data, it seems that they occur in waves. For the anomalies, there doesn't seem to be a particular peak time; they just occur throughout the entire timespan.

It does appear that that they have defined spikes near the start of the first transaction, and that some of the spikes do occur where normal transactions are in the "trough" of the wave pattern shown. However, a good portion of the fraudulent transactions still occur where normal transactions are at a maximum.

So what does the data for the other columns look like? Let's look at some interesting plots for V1:

```
plot_histogram(df, bins, "V1")
```

Refer to Figure 1-19.

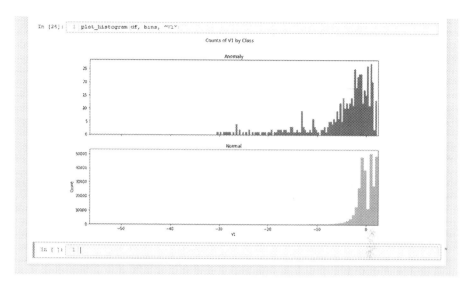

Figure 1-19. *Using the plot_histogram() function to plot the data in the column V1 in df*

Here, you can see a clear difference in the distribution of points for each class over the same V1 values. The range of values that the fraudulent transactions encompass extend well into the values for V1. Let's keep exploring, looking at how the values for Amount relate to V1:

```
plot_scatter(df, "Amount", "V1", sharey=True)
```

What the sharey parameter does is it forces both subplots to share the same y-axis, meaning the plots are displayed on the same scale. You are specifying this so it will be easier to tell what the distribution of the anomalous points looks like in comparison to the normal points. Refer to Figure 1-20.

27

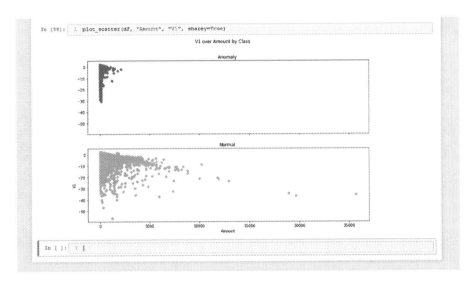

Figure 1-20. *Using the plot_scatter() function to plot the values in the columns Amount on the x-axis and V1 on the y-axis in df*

From this graph, the fraudulent points don't seem out of place compared to all of the other normal points.

Let's continue and look at how time relates to the values for V1:

```
plot_scatter(df, "Time", "V1", sharey=True)
```

Refer to Figure 1-21.

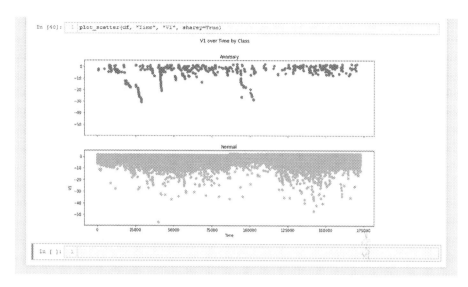

Figure 1-21. *Using the plot_scatter() function to plot the values in the columns Time on the x-axis and V1 on the y-axis in df*

Other than a few defined spikes that stand out from where the normal points would have been, most of the fraudulent data in this context seems to blend in with the normal data.

Doing this one at a time for all of the other values will be tedious, so let's just plot them all at once using a simple script. Here is the code to plot all of the frequency counts for each column from V1 to V28:

```
for f in range(1, 29):
    print(f'V{f} Counts')
    plot_histogram(df, bins, f'V{f}')
```

Refer to Figure 1-22.

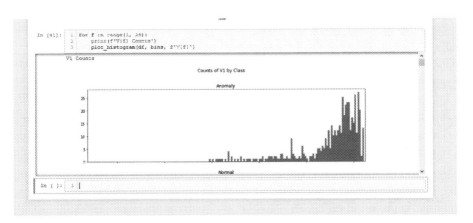

Figure 1-22. *A script to plot histograms using the plot_histogram()*
function for data in each column from V1 to V28 in df

Since the output has been minimized, hover where the bar darkens
and click to expand the output so you can see the graphs a lot better. Refer
to Figure 1-23.

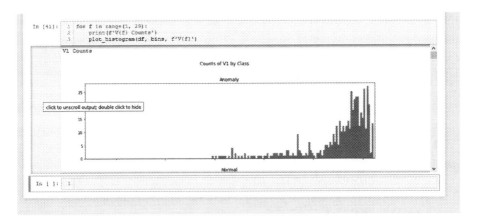

Figure 1-23. *Hovering over the bar to the left of the plots (it should*
darken and show the tooltip as shown) and clicking it to expand the
output

Now you should see something like in Figure 1-24.

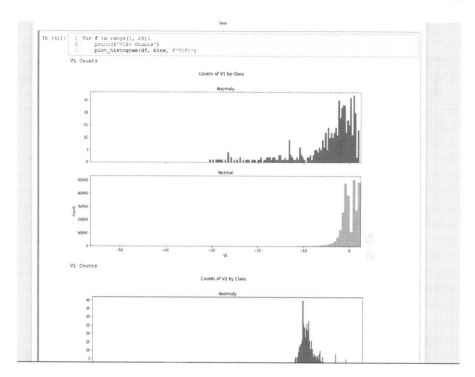

Figure 1-24. *What the expanded output should look like. All of the graphs should display continuously, as depicted in the figure*

Scrolling through, you can see a lot of interesting graphs such as Figure 1-25 and Figure 1-26.

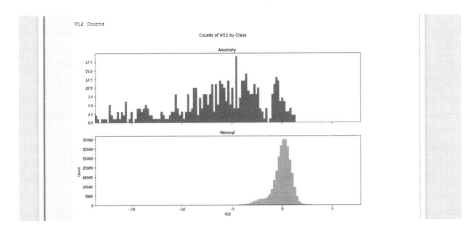

Figure 1-25. *A histogram of data for the column V12 in df. As you can see, there is a very clear deviation seen with the anomalous values compared to the normal values. Both plots share the same x-axis scale, so while the counts might be very low compared to the normal values, they are still spread out far more than the normal values for the same range of V12 column values*

In this case, you can see a clear differentiation between the fraudulent data and the normal data that you didn't see in the graphs earlier. And so, features such as V12 are certainly more important in helping give the model a better context.

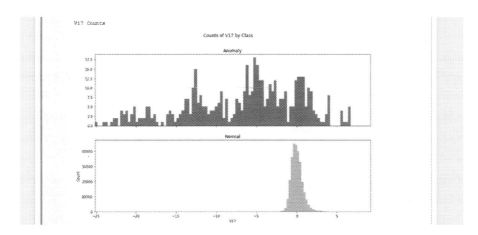

Figure 1-26. *A histogram of data for the column V17 in df. Just like with the column V12, there is also a clear deviation seen with the anomalous values compared to the normal values. This indicates that the column V17 is more likely to help the model learn how to differentiate between normal and fraudulent transactions than some of the other columns that don't show such a deviance*

This time you can see an even bigger difference between fraudulent data and normal data. Once again, it's features like V12 and V17 that hold the data that will help the model understand how to differentiate between the anomalies and the normal points.

To minimize the output, click the same bar as earlier when you expanded the output. Let's now look at how all of these data points vary according to time:

```
for f in range(1, 29):
    print(f'V{f} vs Time')
    plot_scatter(df, "Time", f'V{f}', sharey=True)
```

Once again, expand the output and explore the graphs. Refer to Figure 1-27 and Figure 1-28 to see some interesting results.

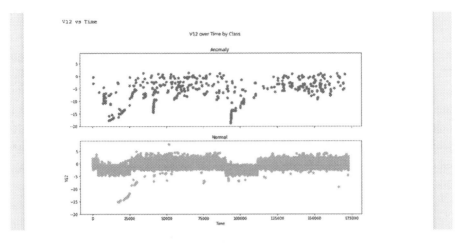

Figure 1-27. *The scatterplot for Time on the x-axis and V12 on the y-axis shows a deviation between the anomalies and the normal data points. Although a significant portion of the anomalies fall under the band of normal points, there are still a good number of anomalies that fall out of that range. And so you can see that against Time, the data for the column V12 also shows this deviation from the normal data points*

Once again, with V12 you can see a significant difference between the anomalies and the normal data points. A good portion of the anomalies remain hidden within the normal data points, but a significant amount of them can be differentiated from the rest.

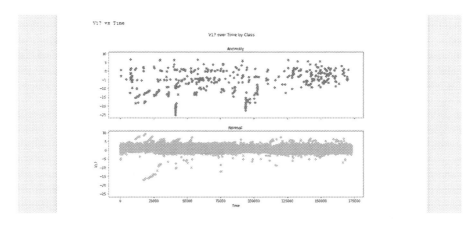

Figure 1-28. *The scatterplot for Time on the x-axis and V17 on the y-axis shows a deviation between the anomalies and the normal data points. As with the values for V12, you can observe another deviation between the normal points and the fraudulent points. In this case, the difference seems to be a bit more pronounced, as the anomalies seem to be more spread out than in Figure 1-27*

The difference between the anomalies and the normal points are highlighted even further when looking at V17. It seems that even in relation to time, columns V12 and V17 hold data that best help distinguish fraudulent transactions from normal transactions. You can see in the graph that a few normal points are with the anomalous points as well, but hopefully the model can learn the true difference taking into account all of the data.

Finally, let's see the relationship between each of these columns and Amount:

```
for f in range(1, 29):
    print(f'Amount vs V{f}')
    plot_scatter(df, f'V{f}', "Amount", sharey=True)
```

This time there seems to be a few more graphs more clearly showing the differences between the normal and fraudulent points. Refer to Figure 1-29, Figure 1-30, and Figure 1-31.

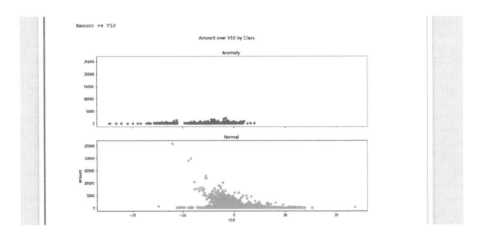

Figure 1-29. *Looking at the scatterplot for Amount on the y-axis and V10 on the x-axis, you can see a pronounced deviation of fraudulent points from the normal points. For the relationship of the V columns against Amount, it seems that more columns show an increased deviation compared to the earlier plots. This difference is not so large, as you still see that a sizeable portion of the anomalies are within the normal data cluster. However, this still gives the model some context in how a fraudulent transaction differs from a normal transaction*

The graphs from V9 through V12 all show a clear differentiation between the anomalies and the normal points, even if a good portion of the anomalies are within the cluster of normal points. One thing to note is that it may not be the same anomalies that differ each time in the graphs, allowing the model to better learn how to differentiate between the anomalies and the normal points.

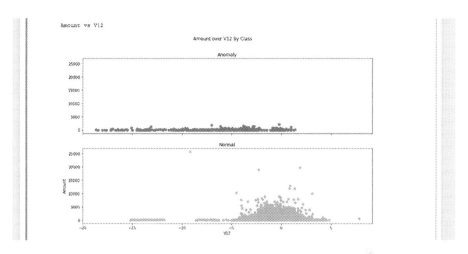

Figure 1-30. *A scatterplot for the column Amount on the y-axis and V12 on the x-axis. Once again, you can see a pronounced deviation of fraudulent points from the normal points. In this case, the majority of fraudulent points seem to deviate from the normal point cluster. You can also see that there is a band of normal points far from the main cluster, and that the band coincides with the anomalous data points. It is a possible reason to keep in mind if the model classifies points like these as anomalies*

You can once again see that V12 consistently differentiates between anomalies and normal data. However, there is still the problem of a good portion of the anomalies staying hidden within the normal data cluster.

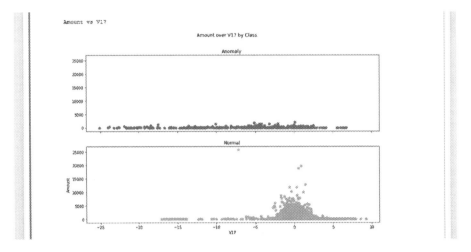

Figure 1-31. *A scatterplot for the column Amount on the y-axis and V17 on the x-axis. Just as with Figure 1-30, you can see a deviation again of fraudulent points from the normal point cluster. Once again, the majority of fraudulent points show this deviation, but you can also see some normal points that coincide with these anomalous points*

You can also see that this differentiation between normal points and fraudulent points holds for V17 looking at transaction amounts.

You could also look at the data for each of the V columns and plot them against each other, but that's more useful to help identify precise changes in trends that will be more useful to know if you want to further train the model to improve its performance on the new data. First of all, it's possible that not every feature is very significant. So, if trends do shift, it does not necessarily mean that the model's performance will be downgraded.

Thorough analysis of the data helps data scientists get a much better understanding of how the various data columns relate to each other and lets them identify if trends are shifting over time. As data is continuously collected over time, data biases and trends are bound to shift. So perhaps a year from now, it's the column V18 that shows profound differences between anomalous points and normal points, and V17 now shows that most anomalous points are contained within the cluster of normal points.

Summary

Data analysis is a crucial step in the process of creating a machine learning solution. Not only does it determine the type of model and influence the set of features that will be selected for the training process, but it also helps identify any changes in trends over time that may signify that the model needs to be further trained. You explored and analyzed the data in the credit card data set, generated many plots to get an idea of the relationship between the two plotted variables, and identified some features that distinguish between normal points and anomalies. In the next chapter, you will process the data to create various subsets to help train several types of machine learning models.

CHAPTER 2

Building Models

In this chapter, we will go over how to build a simple logistic regression model in both scikit-learn and PySpark. We will also go over the process of k-fold cross validation to tune a hyperparameter in scikit-learn.

Introduction

In the previous chapter, you loaded the credit card data set and analyzed the distribution of its data. You also looked at the relationships between the features and got a general idea of how heavily they influence the labels.

Now that you've gained a better understanding of the data set, you will proceed with building the models themselves. You will be using the same credit card data set as in the previous chapter. In this chapter, you will look at two frameworks: **scikit-learn**, and **PySpark**. The models you build in scikit-learn and in PySpark will stay relevant for the rest of the book, as you will be using both of them later on when you host them on cloud services to make predictions. You will keep it simple and construct logistic regression models in these two frameworks. Since the input data format is different for these two frameworks, you can't just conduct the data processing in advance and use those train/test/validate sets for these two frameworks. However, it is possible to do so for scikit-learn and Keras, for example, depending on how the last layer is constructed in the Keras model.

© Sridhar Alla, Suman Kalyan Adari 2021
S. Alla and S. K. Adari, *Beginning MLOps with MLFlow*,
https://doi.org/10.1007/978-1-4842-6549-9_2

You will be performing the validation step with the scikit-learn model to tune a hyperparameter. **Hyperparameters** can be thought of as model-related parameters that influence the training process and result.

That being said, let's get started with scikit-learn and build a logistic regression model. One thing to note is that we will provide a lot of commentary in the scikit-learn model that we may skip over in the PySpark example, so be sure to at least read through the process for scikit-learn to get a general idea of how train-test-validate works.

Scikit-Learn

Before we get started, here are the packages and their versions that you will need. We will provide an easy way for you to check the versions of your packages within the code itself.

Here are the versions of our configuration:

- Python 3.6.5

- numpy 1.18.5

- pandas 1.1.0

- matplotlib 3.2.1

- seaborn 0.10.1

- sklearn 0.22.1.post1

In the code below, you will find that some of the imports are unnecessary, such as importing all of sklearn when you only use a bit of its functionality. This is done for the purpose of displaying the version and such statements have a # beside them.

Data Processing

So now, let's begin with the import statements:

```
import numpy as np
import pandas as pd
import matplotlib #
import matplotlib.pyplot as plt
import seaborn as sns
import sklearn #
from sklearn.linear_model import LogisticRegression
from sklearn.model_selection import train_test_split
from sklearn.preprocessing import StandardScaler
from sklearn.metrics import roc_auc_score, plot_roc_curve,
confusion_matrix
from sklearn.model_selection import KFold

print("numpy: {}".format(np.__version__))
print("pandas: {}".format(pd.__version__))
print("matplotlib: {}".format(matplotlib.__version__))
print("seaborn: {}".format(sns.__version__))
print("sklearn: {}".format(sklearn.__version__))
```

Refer to Figure 2-1 to see the output.

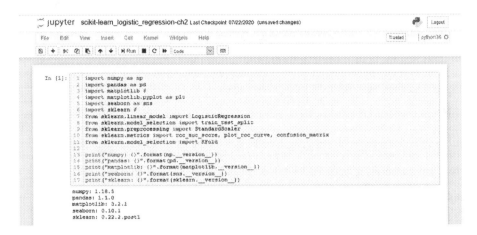

Figure 2-1. *The output showing the printed versions of the modules you will need. Some modules are imported for the sake of printing the versions and have been marked with a # beside them to indicate that they are not necessary to run the code*

Now you can move on to loading the data. You will be using the same credit card dataset as from the previous chapter:

```
data_path = "data/creditcard.csv"
```

```
df = pd.read_csv(data_path)
```

Refer to Figure 2-2 to see this code in a cell.

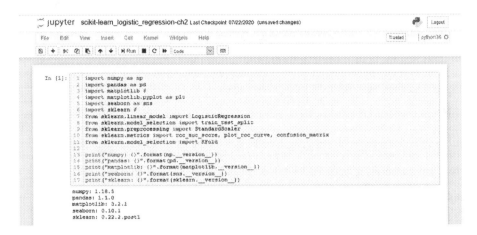

Figure 2-2. *Loading the data frame using pandas. The credit card data set is located in a folder called* data, *which is located in the same directory as the notebook file*

There shouldn't be any output from loading the data frame. To see the data frame you just loaded, call the following to ensure it has read the data correctly:

```
df.head()
```

You should see something like in Figure 2-3.

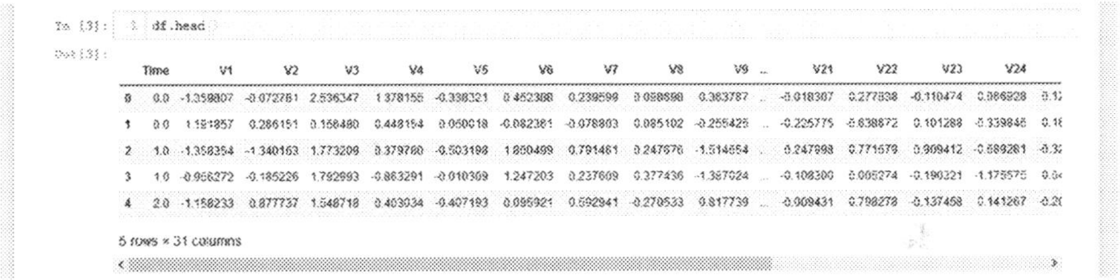

In [3]: df.head()

Out[3]:

	Time	V1	V2	V3	V4	V5	V6	V7	V8	V9	...	V21	V22	V23	V24	
0	0.0	-1.359807	-0.072781	2.536347	1.378155	-0.338321	0.462388	0.239599	0.098698	0.363787	...	-0.018307	0.277838	-0.110474	0.066928	0.1
1	0.0	1.191857	0.266151	0.166480	0.448154	0.060018	-0.082361	-0.078803	0.085102	-0.255425	...	-0.225775	-0.638672	0.101288	-0.339846	0.1
2	1.0	-1.358354	-1.340163	1.773209	0.379780	-0.503198	1.800499	0.791461	0.247676	-1.514654	...	0.247998	0.771679	0.909412	-0.689281	-0.3
3	1.0	-0.966272	-0.185226	1.792993	-0.863291	-0.010309	1.247203	0.237609	0.377436	-1.387024	...	-0.108300	0.005274	-0.190321	-1.175575	0.6
4	2.0	-1.158233	0.877737	1.548718	0.403034	-0.407193	0.095921	0.592941	-0.270533	0.817739	...	-0.009431	0.798278	-0.137458	0.141267	-0.2

5 rows × 31 columns

Figure 2-3. *The output of the head() function. The data has loaded correctly, and you can see the first five rows of the data frame*

If you remember from the previous chapter, there is a massive imbalance in the distribution of data between the normal data and the anomalies. Because of this, you are going to take a slightly alternative approach in how you craft this data.

This is where data analysis comes into play. Because you know that a massive disparity between the data counts in each class exists, you will now take care to specially craft the data sets so that it is ensured that a good amount of anomalies end up in each data set. If you simply select 100,000 data points from `df`, split it into your training/test/validate sets and continue, it is entirely possible that very few or even no anomalies end up in one or more of those sets. At that point, you would have a lot of trouble in getting the model to properly learn this task.

This is why you will be splitting up the anomalies and normal points to create your training/test/validate sets.

With that in mind, let's create data frames for the normal points and for the fraudulent points:

```
normal = df[df.Class == 0].sample(frac=0.5, random_state=2020).
reset_index(drop=True)
anomaly = df[df.Class == 1]
```

You have set the random_state to a specific value so that the results of the random sampling should be the same no matter how many times you repeat it, helping with reproducibility. Unfortunately, given the nature of how models learn, you cannot expect to get the same results every time for something like neural networks, for example.

In the code, you filter out the respective values by class, and sample 50% of the entire data frame's normal points to comprise the normal data in this context.

Refer to Figure 2-4 to see this code in a cell.

Figure 2-4. *Filtering the data frame values by class to create the normal and anomaly data frames. The normal data frame contains 50% of all normal data points, randomly selected as determined by the seed (random_state)*

You can add some code to check the shapes as well:

```
print(f"Normal: {normal.shape}")
print(f"Anomaly: {anomaly.shape}")
```

Refer to Figure 2-5 for the output.

Figure 2-5. *Printing the shapes of the normal and anomaly data frames. There is a clear difference in the number of entries in the two data frames*

As you can see, there is still a big disparity between the normal points and the anomalies. In the case of logistic regression, the model is still able to learn how to distinguish between the two, but in the case of neural networks, for example, this disparity means the model never really learns how to classify anomalies. However, as you will see later in this chapter, you can tell the model to weigh the anomalies far more in its learning process compared to the normal points.

Now you can start creating the train/test/validate split. However, scikit-learn provides functionality to create train/test splits only. To get around that, you will create train and test sets, and then split the train set again into train and validate sets.

First, you will split the data into train and test data, keeping the normal points and anomalies separate. To do this, you will use the `train_test_split()` function from scikit-learn. Commonly passed parameters are

- x: The x set you want to split up

- y: The y set you want to split up corresponding to the x set

- `test_size`: The proportion of data in x and y that you want to randomly sample for the test set.

And so, to split up x and y into your training and testing sets, you may see code like the following:

```
x_train, x_test, y_train, y_test = train_test_split(x, y, test_size=0.2, random_state = 2020)
```

Just like earlier, `random_state` is setting the random seed so that every time you run it, the data will be split the same way.

If you don't pass in the y parameter, you simply get a split on the x data. And so, keeping that in mind, let's split up your normal points and anomalies into training and testing sets:

```
normal_train, normal_test = train_test_split(normal, test_size
= 0.2, random_state = 2020)
anomaly_train, anomaly_test = train_test_split(anomaly,
test_size = 0.2, random_state = 2020)
```

There should be no output but refer to Figure 2-6 to see the code in a cell.

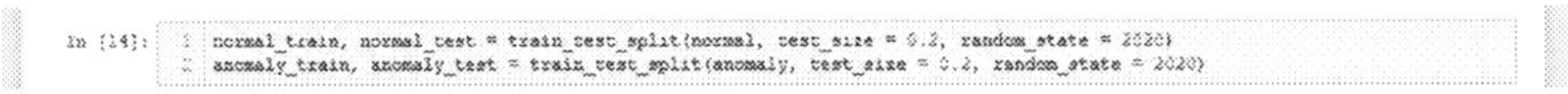

Figure 2-6. *Splitting the normal and anomaly data frames into train and test subsets. The respective test sets comprise 20% of the original sets*

Now, you can create your training and validation sets by calling the same function on the respective training sets. You don't want to split it by 20% again, though, since the training set is already 80% of the original data set. If you used a 20% split again, the validation set would be 16% of the original data, and the training set would be 64% of the original data. You will instead be doing a 60-20-20 split for the training, testing, and validation data, respectively, and so you will be using a new test_size value of 0.25 to ensure these proportions hold (0.25 * 0.8 = 0.2).

With that in mind, let's create your training and validation splits:

```
normal_train, normal_validate = train_test_split(normal_train,
test_size = 0.25, random_state = 2020)
anomaly_train, anomaly_validate = train_test_split(anomaly_
train, test_size = 0.25, random_state = 2020)
```

Refer to Figure 2-7 to see the code in a cell.

```
In [30]:    1  normal_train, normal_validate = train_test_split(normal_train, test_size = 0.25, random_state = 2020)
            2  anomaly_train, anomaly_validate = train_test_split(anomaly_train, test_size = 0.25, random_state = 2020)
```

Figure 2-7. *You create train and validate splits from the training data. You have chosen to make the validation set comprise 25% of the respective original training sets. As these original training sets themselves comprise of 80% of the original normal and anomaly data frames, the respective validation splits are 20% (0. 25 * 0.8) of their original normal and anomaly data frames. And so, the final training split also becomes 60% of the original, as 0.75 * 0.8 = 0.6*

To create your final training, testing, and validation sets, you have to concatenate the respective normal and anomaly data splits.

First, you define x_train, x_test, and x_validate:

```
x_train = pd.concat((normal_train, anomaly_train))
x_test = pd.concat((normal_test, anomaly_test))
x_validate = pd.concat((normal_validate, anomaly_validate))
```

Next, you define y_train, y_test, and y_validate:

```
y_train = np.array(x_train["Class"])
y_test = np.array(x_test["Class"])
y_validate = np.array(x_validate["Class"])
```

Finally, you have to drop the column Class in the x sets since it would defeat the purpose of teaching the model how to learn what makes up a normal and a fraudulent transaction if you gave it the label directly:

```
x_train = x_train.drop("Class", axis=1)
x_test = x_test.drop("Class", axis=1)
x_validate = x_validate.drop("Class", axis=1)
```

To see all this code in a cell, refer to Figure 2-8.

```
In [8]:  1  x_train = pd.concat((normal_train, anomaly_train))
         2  x_test = pd.concat((normal_test, anomaly_test))
         3  x_validate = pd.concat((normal_validate, anomaly_validate))
         4
         5  y_train = x_train["Class"]
         6  y_test = x_test["Class"]
         7  y_validate = x_validate["Class"]
         8
         9  x_train = x_train.drop("Class", axis=1)
        10  x_test = x_test.drop("Class", axis=1)
        11  x_validate = x_validate.drop("Class", axis=1)
```

Figure 2-8. *Creating the respective x and y splits of the training, testing, and validation sets. The x sets are the combinations of the normal and anomaly sets for each split (train, test, validate), while the y sets are simply the data in the Class columns of those x sets. You then drop the label column from the x sets*

Let's get the shapes of the sets you just created:

```
print("Training sets:\nx_train: {} y_train: {}".format(x_train.
shape, y_train.shape))
print("\nTesting sets:\nx_test: {} y_test: {}".format(x_test.
shape, y_test.shape))
print("\nValidation sets:\nx_validate: {} y_validate:
{}".format(x_validate.shape, y_validate.shape))
```

Refer to Figure 2-9 to see the output.

```
In [32]:  1  print("Training sets:\nx_train: {} y_train: {}".format(x_train.shape, y_train.shape))
          2  print("\nTesting sets:\nx_test: {} y_test: {}".format(x_test.shape, y_test.shape))
          3  print("\nValidation sets:\nx_validate: {} y_validate: {}".format(x_validate.shape, y_validate.shape))

Training sets:
x_train: (85588, 30) y_train: (85588,)

Testing sets:
x_test: (28531, 30) y_test: (28531,)

Validation sets:
x_validate: (28531, 30) y_validate: (28531,)
```

Figure 2-9. *Printing the output of the different sets. The three sets should comprise 60%, 20%, and 20% of the original union of the normal and anomaly sets*

Looking at the data analysis, you can see that some of the values get really large. The fine details are beyond the scope of this book, but when some features have a relatively small range but others have an extremely large range (think of the range of V1 and Time from the previous chapter), the model will have a much harder time learning.

In more detail, the model will have a hard time optimizing the cost function and may take many more steps to converge, if it is able to do so at all.

And so it is better to scale everything down by normalizing the data. You will be using scikit-learn's `StandardScaler`, which normalizes all of the data such that the mean is 0 and the standard deviation is 1.

Here is the code to standardize your data:

```
scaler = StandardScaler()
scaler.fit(pd.concat((normal, anomaly)).drop("Class", axis=1))

x_train = scaler.transform(x_train)
x_test = scaler.transform(x_test)
x_validate = scaler.transform(x_validate)
```

It is important to note that you are fitting the scaler on the entire data frame so that it standardizes all of your data in the same way. This is to ensure the best results since you don't want to standardize x_train, x_test, and x_validate in their own ways since it would create discrepancies in the data and would be problematic for the model. Of course, once you've deployed the model and start receiving new data, you would still standardize it using the scaler from the training process, but this new data could possibly come from a slightly different distribution than your training data. This would especially be the case if trends start shifting - this new standardized data could possibly lead to a tougher time for the model since it wouldn't fit very well in the distribution that the model trained on.

Refer to Figure 2-10 to see the code in a cell.

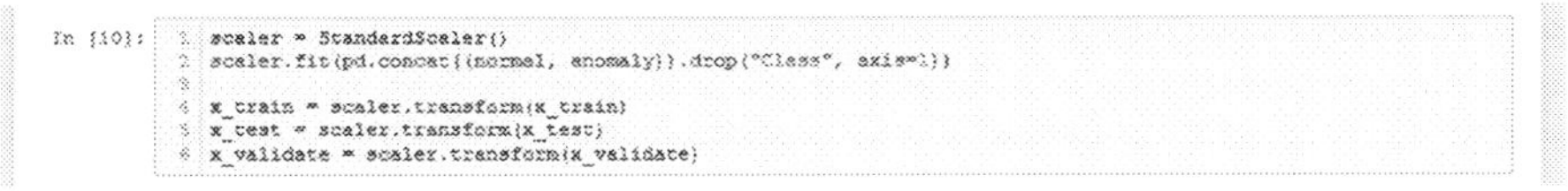

Figure 2-10. *Fitting a standard scaler object on a concatenation of the normal and anomaly data frames. This is done so that each of the train, test, and validate subsets will be scaled according to the same standards, ensuring that there are no discrepancies between the scaling of the data*

Model Training

Finally, you can now define your logistic regression model:

```
sk_model = LogisticRegression(random_state=None, max_iter=400,
solver='newton-cg').fit(x_train, y_train)
```

Refer to Figure 2-11 to see the code in a cell. There should not be any outputs after execution if it all goes well. Any errors you might see could involve a failure to converge. For that, changing the max_iter parameter could help, and changing the solver algorithm could help as well.

Figure 2-11. *Defining the logistic regression model and training it on the training data*

After the training process, either the evaluation step or validation step can come next. As long as the testing set and the validation set come from different distributions (the validation set is derived from the training set, while the testing set is derived from the original data), the model is technically seeing new data in the evaluation and in the validation processes.

The context also matters. If you are using the validation process to select the best model out of a set of trained models, then the validation process can come after the training process. You can still evaluate one or all of your trained models, but it could be unnecessary because in this context you're trying to find the best model for the code.

In the context where you're trying to tune your hyperparameters for a model you are going to stick with, it doesn't matter whether you do the evaluation first or the validation first. Doing the evaluation first, as you will be doing shortly, can give you a good idea of how well the model is doing currently before starting the validation step. The model will never learn from the evaluation data, so there's no harm in evaluating the model on this data.

In this example, you are looking at tuning the hyperparameter for class weights (how much to weight a normal sample and how much to weight a fraudulent sample).

But first, let's evaluate your model to get a deeper understanding of how everything works.

Model Evaluation

You can now look at accuracy and AUC scores. First, you find the accuracy using the built-in score function of the model:

```
eval_acc = sk_model.score(x_test, y_test)
```

Next, let's get the list of predictions from the model to help calculate the AUC score. AUC is usually a better metric since it better explains the performance of the model. The general gist of it is that a model that perfectly classifies every point correctly will have an AUC score of 100%.

The problem with accuracy in this context is that if there are 100,000 normal points and perhaps around 100 anomalies, the model can classify all of the normal points correctly and none of the anomalies and still get a really high accuracy above 99%. However, the AUC score would show a value much lower at around 0.5. An AUC of 0.5 means that the model knows nothing and is practically just guessing randomly, but in this case, it means the model only ever predicts "normal" for any point it sees. In other words, it hasn't actually learned much of anything if it doesn't know how to predict an anomaly.

It's also worth mentioning that AUC isn't the sole metric by which one should base the worthiness of a model, since context matters. In this case, normal points far outnumber anomalies, so accuracy is a relatively poor metric to solely judge model performance on. AUC scores in this case would reflect the mode's performance well, but it's also possible to get higher AUC scores but lower accuracy scores. That just means you must look at the results carefully to understand exactly what's happening. To help with this, you will look at a "confusion matrix" shortly.

Now, let's get the predictions and calculate the AUC score:

```
preds = sk_model.predict(x_test)
auc_score = roc_auc_score(y_test, preds)
```

Finally, let's print out the scores:

```
print(f"Auc Score: {auc_score:.3%}")
print(f"Eval Accuracy: {eval_acc:.3%}")
```

Refer to Figure 2-12 to see all three of the cells above and the output that results.

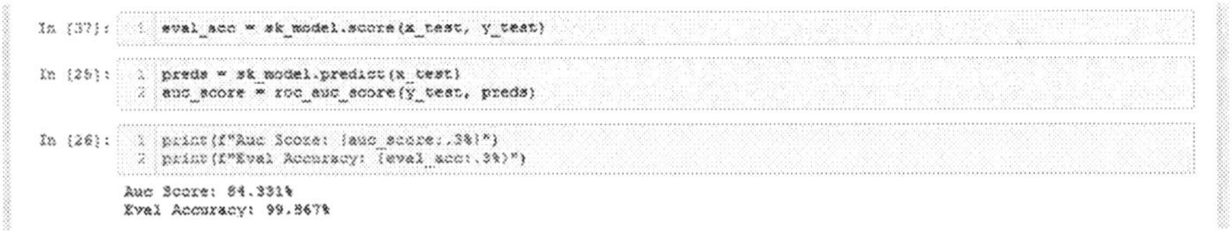

***Figure 2-12.** Printing out the AUC score and the accuracy for the scikit-learn logistic regression model*

In this case, both the AUC score and the accuracy score are high. Between the two, the accuracy score is definitely inflated by the number of normal points that exist, but the AUC score indicates that the model does a pretty good job at distinguishing between the anomalies and the normal points.

Scikit-learn actually provides a function that lets you see the ROC curve—the figure from which the AUC score (or "area under curve") is derived from. Run the following:

```
roc_plot = plot_roc_curve(sk_model, x_test, y_test,
name='Scikit-learn ROC Curve')
```

Refer to Figure 2-13 for the output.

Figure 2-13. *The ROC curve generated for the logistic regression model you just trained. An ROC curve starting with a true positive value of 1.0 at a false positive value of 0.0 is the best possible curve in theory. From that point, it should keep going right while maintaining its value as it hits 1.0 on the x-axis. This graph is quite close to that ideal, hence why the AUC score is so high at 0.98. The discrepancy in AUC score here compared to when you calculated it earlier has to do with how the value is actually calculated*

What's basically happening is that scikit-learn takes in the model and the evaluation set to dynamically generate the curve as it predicts on the test sets. The metrics you see on the axes are derived from how correctly the model predicts each of the values. The "true positive rate" and the "false positive rate" are derived from the values on the confusion matrix that you will see below.

From that graph, the AUC score is generated. You can see that it differs from the score that was calculated earlier, but this can be attributed to the two functions calculating the scores slightly differently.

Let's now build the confusion matrix and plot it using seaborn:

```
conf_matrix = confusion_matrix(y_test, preds)
ax = sns.heatmap(conf_matrix, annot=True, fmt='g')
ax.invert_xaxis()
ax.invert_yaxis()
plt.ylabel('Actual')
plt.xlabel('Predicted')
```

Refer to Figure 2-14 for the output.

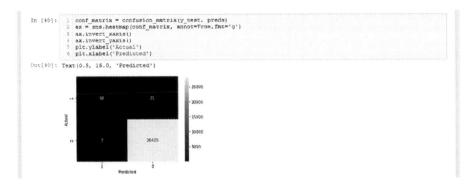

Figure 2-14. *The confusion matrix plot of the results of training. The accuracy for the normal points is very good, but the accuracy for the anomaly points is ok. There is still further room for improvement looking at these results, as you have not tuned the hyperparameters of the model yet, but it already does ok in detecting anomalies. The goal now is to keep the accuracy for the normal points as high as possible, or at a high enough level that's acceptable, while raising the accuracy for the anomaly points as high as possible. Based on this confusion matrix plot, you can now see that the lower AUC score is more accurate at reflecting the true performance of the model. You can see that a non-negligible amount of anomalies were falsely classified as normal, hence an AUC score of 0.84 is a much better indicator of the model's performance than the graph's apparent score of 0.98*

This is what a confusion matrix looks like. The y-axis consists of the true labels, while the x-axis consists of predicted labels. When the true label is "0" and the model predicts "0," we call this a **true negative**. "True" refers to the true label, and "negative" refers to the label the model predicts.

What counts as "positive" and what counts as "negative" can differ. In tasks such as disease detection, if a test finds someone to have the disease, they are said to "test positive." Otherwise, they "test negative." Anomaly detection is similar. When a model thinks that a point is an anomaly, it flags it with the label "1." And so, a point is labeled "positive" if the model thinks it is an anomaly, and "negative" if it doesn't.

You may notice that we have inverted the axes in the code. This is simply to get it in the format so that the top left of the matrix corresponds to "true positives," the top right of the matrix corresponds to "false negatives," the bottom left of the matrix corresponds to "false positives," and the bottom right of the matrix corresponds to "true negatives."

To quickly recap these concepts:

- **True positives** are values that the model predicts as positive that actually are positive.

- **False negatives** are values that the model predicts as negative that actually are positive.

- **False positives** are values that the model predicts as positive that actually are negative.

- **True negatives** are values that the model predicts as negative that actually are negative.

To look at how well the model identifies anomalies, look at the 1 row on the y-axis. The sum of this row should equal the total number of anomalies in the test set: 99 anomalies. The model predicted about 68.7% of the anomalies correctly (68/(68+31)) and predicted 99.98% of the normal points correctly (28425/(28425 + 7)) looking at the bottom row.

As you can see, the confusion matrix gives us a really good look at the true performance of the model. You now know that it does very well in the task of predicting normal points but does an ok job at predicting anomalies. That being said, the model can still predict a majority of anomalies correctly. And so you can see that the AUC score of 0.84 was much more accurate at indicating the performance of the model than the graph, which had an AUC of 0.98. With an AUC of 0.98, you can expect that there are very, very few instances of false negatives or false positives.

Model Validation

Let's now look at how to use the process of k-fold cross-validation to compare several hyperparameter values. After the validation process has ended, you will compare the evaluation metrics to get a better idea of what hyperparameter setting works best.

The hyperparameter you want to tune is how much you want to weight the anomalies by compared to the normal data points. By default, both of them are weighted equally. Let's define a list of weights to iterate over:

```
anomaly_weights = [1, 5, 10, 15]
```

Next, you define the number of folds and initialize your data fold generator:

```
num_folds = 5
kfold = KFold(n_splits=num_folds, shuffle=True,
random_state=2020)
```

What this KFold() function does is that it splits the data passed in into num_folds different partitions. A single fold acts as a validation set at a time, while the rest of the folds are used for training. In this context, the "validation fold" is basically what the model will be evaluating on. It is called "validation" since it helps us get an idea of how the model is doing on data it has never seen before.

If you have built deep learning models before, you may know that during the training process, you can split a small portion of the training set aside as a validation set. This lets you know during training if you're overfitting or not, as decreasing training loss and increasing validation loss would indicate.

Refer to Figure 2-15 to see the code above in cells.

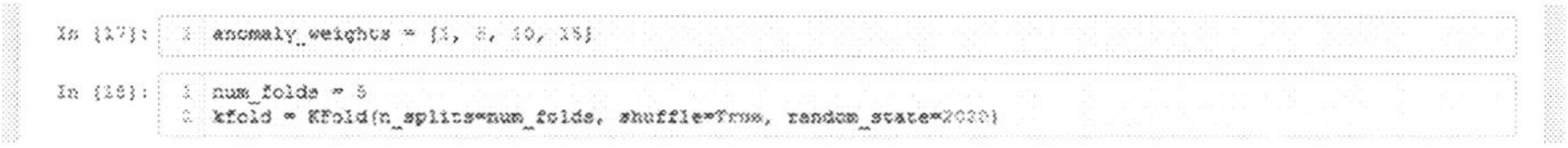

```
In [17]:   1  anomaly_weights = [1, 5, 10, 15]
In [18]:   1  num_folds = 5
           2  kfold = KFold(n_splits=num_folds, shuffle=True, random_state=2020)
```

Figure 2-15. *Setting the different values for anomaly weights to test with the validation script and constructing the KFold data generator. In this case, you are using five folds, so the data passed in will be split five ways*

Now you define the validation script:

```
logs = []

for f in range(len(anomaly_weights)):
    fold = 1
    accuracies = []
    auc_scores= []
    for train, test in kfold.split(x_validate, y_validate):

        weight = anomaly_weights[f]

        class_weights= {
            0:1,
            1: weight
        }

        sk_model = LogisticRegression(random_state=None,
                                      max_iter=400,
                                      solver='newton-cg',
                                      class_weight=class_
                                      weights).fit(x_
                                      validate[train],
                                      y_validate[train])
```

```
    for h in range(40): print('-', end="")
    print(f"\nfold {fold}\nAnomaly Weight: {weight}")

    eval_acc = sk_model.score(x_validate[test],
    y_validate[test])
    preds = sk_model.predict(x_validate[test])

    try:
        auc_score = roc_auc_score(y_validate[test], preds)
    except:
        auc_score = -1

    print("AUC: {}\neval_acc: {}".format(auc_score, eval_acc))

    accuracies.append(eval_acc)
    auc_scores.append(auc_score)

    log = [sk_model, x_validate[test], y_validate[test], preds]
    logs.append(log)

    fold = fold + 1
print("\nAverages: ")
print("Accuracy: ", np.mean(accuracies))
print("AUC: ", np.mean(auc_scores))

print("Best: ")
print("Accuracy: ", np.max(accuracies))
print("AUC: ", np.max(auc_scores))
```

That's a lot to take in at once, so be sure to refer to Figure 2-16 to make sure your code is formatted correctly.

```
In [33]:    1  logs = {}
            2
            3  for f in range(len(anomaly_weights)):
            4      fold = 1
            5      accuracies = []
            6      auc_scores= []
            7      for train, test in kfold.split(x_validate, y_validate):
            8
            9          weight = anomaly_weights[f]
           10
           11          class_weights= {
           12              0:1,
           13              1: weight
           14          }
           15
           16          sk_model = LogisticRegression(random_state=None,
           17                                        max_iter=400,
           18                                        solver='newton-cg',
           19                                        class_weight=class_weights).fit(x_validate[train], y_validate[train])
           20
           21          for h in range(40): print('-', end="")
           22          print(f"\nfold {fold}\nAnomaly Weight: {weight}")
           23
           24
           25          eval_acc = sk_model.score(x_validate[test], y_validate[test])
           26          preds = sk_model.predict(x_validate[test])
           27
           28          try:
           29              auc_score = roc_auc_score(y_validate[test], preds)
           30          except:
           31              auc_score = -1
           32
           33
           34          print("AUC: {}\neval_acc: {}".format(auc_score, eval_acc))
           35
           36          accuracies.append(eval_acc)
           37          auc_scores.append(auc_score)
           38
           39          log = [sk_model, x_validate[test], y_validate[test], preds]
           40          logs.append(log)
           41
           42          fold = fold + 1
           43
           44      print("\nAverages: ")
           45      print("Accuracy: ", np.mean(accuracies))
           46      print("AUC: ", np.mean(auc_scores))
           47
           48      print("Best: ")
           49      print("Accuracy: ", np.max(accuracies))
           50      print("AUC: ", np.max(auc_scores))
           51
```

Figure 2-16. *The validation script in a cell. The script is quite long, so be sure it is formatted correctly because a single space misalignment can cause issues*

Before you run the script, let's go over what the code does, as that was a lot of code thrown out at once.

The first loop goes over each of the anomaly weights. You set the fold number here equal to 1 and define empty lists to hold values for accuracy and AUC scores for each run with the current weight parameter.

The second loop goes over the five fold boundaries that the KFold() object defines. You set the class_weights dictionary and pass it into the model as a hyperparameter. After the training process, you evaluate as usual. There is a try-except block for the AUC score in the event that the

61

fold generated only has values of one class (so really if it only has normal data and no anomalies). If the AUC score is -1 for any fold, then you know there was a problem with one of the folds.

You do save the model, the validation data, and the predictions so that you can examine the confusion matrix and plot the ROC curve for any run you like. After the end of the five folds, the script then displays averages and the best scores.

The output will be truncated when you run this, so don't forget to expand it like in the previous chapter to look at all of the runs. Feel free to explore the output or even change the number of folds but beware of the results because increasing the number of folds can mean that the number of anomalies must be spread across even more partitions. In this specific context, a lower number of folds is likely to be better because you have so few anomaly points.

When you sift through the output, you can see that the best results occur when the anomaly weight is set to 10. This setting had the highest average AUC score and had the best AUC score as well, resulting in an output like what you see in Figure 2-17.

```
------------------------------------------------
fold 1
Anomaly Weight: 10
AUC: 0.9245604009143662
eval_acc: 0.9985982127212195
------------------------------------------------
fold 2
Anomaly Weight: 10
AUC: 0.9751350672194998
eval_acc: 0.9977216964598669
------------------------------------------------
fold 3
Anomaly Weight: 10
AUC: 0.9313783507133262
eval_acc: 0.9985979670522257
------------------------------------------------
fold 4
Anomaly Weight: 10
AUC: 0.8942972430196292
eval_acc: 0.998422712933754
------------------------------------------------
fold 5
Anomaly Weight: 10
AUC: 0.8820013855427915
eval_acc: 0.9985979670522257

Averages:
Accuracy:   0.9983877112438584
AUC:   0.9214744894819227
Best:
Accuracy:   0.9985982127212195
AUC:   0.9751350672194998
------------------------------------------------
```

Figure 2-17. *Looking at the results of the best setup in the validation script output. The best setup turned out to be one where the anomalies were weighted as 10, as it had the best average AUC score and the best AUC score with the other anomaly weight parameters. The true best weight is likely around an anomaly weight of 10, though you must perform another hyperparemter search with a more narrowed range to find the absolute best setting. You can keep narrowing the search as much as you'd like, but past a certain precision, you will find that you are getting diminishing returns*

Let's examine the plots for this setup since it was the best performer of all of them on average.

First, you load the correct log in the list of logs. Since the anomaly weight was 10, and the second fold performed the best, you want to look at the twelfth index in the entries in logs. (The first five correspond to indices 0-4, and the next five are indices 5-9. With index 10, you begin the first fold with weight ten, so the second fold is at index 11.)

```
sk_model, x_val, y_val, preds = logs[11]
```

Let's look at the ROC curve. Keep in mind that since there is so little data in the validation set, the AUC score may not be so accurate. Here is the code:

```
roc_plot = plot_roc_curve(sk_model, x_val, y_val, name='Scikit-learn ROC Curve')
```

Refer to Figure 2-18 to see the output of the above two cells.

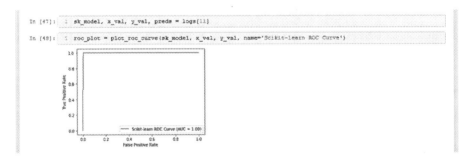

Figure 2-18. *Viewing the ROC curve for a specific validation fold. As you can see, the ROC curve is quite optimal. A perfect ROC curve would start as close as possible to 1.0 on the y-axis while maintaining that level right as it reaches 1.0 on the x-axis. An ROC graph like that would mean the AUC would be as close to 1.0 as possible. In this case, you almost see the perfect AUC curve, and the AUC is stated to be 1.0. The confusion matrix in Figure 2-19 will reveal a lot more about why the AUC score is so low*

This graph looks different compared to the ROC plot you saw earlier. In fact, it almost seems perfect.

Let's look at the confusion matrix to get a better idea of how the model performed on this fold:

```
conf_matrix = confusion_matrix(y_val, preds)
ax = sns.heatmap(conf_matrix, annot=True,fmt='g')
ax.invert_xaxis()
ax.invert_yaxis()
plt.ylabel('Actual')
plt.xlabel('Predicted')
```

The resulting confusion matrix can be seen in Figure 2-19.

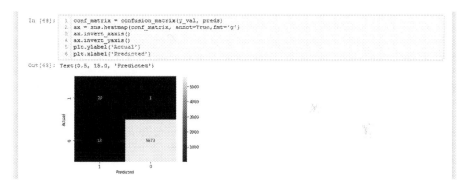

Figure 2-19. *The confusion matrix for a specific validation fold. It has very good accuracy in labeling normal data points and does very well with anomaly points. Additionally, you can see that there are barely any anomalies in this validation fold if you count the entries in the top row: 21 anomalies to 5,685 normal points. It is no wonder, then, that having a higher weight on the anomaly helped the model factor in these anomalies in its learning process, resulting in better performance in anomaly detection*

The model did quite well on correctly classifying the anomalies, but the goal of validation in this case is just to help nudge the hyperparameter setting in the right direction. Based on the results of the validation process, you know that the optimal hyperparameter value lies within the values of 10 and 15 because those two settings produced the best results.

Of course, you can narrow the range further to include values between 10 and 15 for the anomaly weights and repeat this process again and again, further reducing the range until a good, optimal value is found. After a certain precision, however, you will find that you are getting diminishing returns, and that the effort you put into hyperparameter tuning only produces near-negligible boosts in performance.

With that, you now know how to train, evaluate, and validate a logistic regression model in scikit-learn.

PySpark

We have provided the versions of the modules we will be using. Installing PySpark can be a little complicated as it's not a matter of doing `pip install PySpark` depending on the version, so beware of that.

Here are the versions of our configuration:

- Python 3.6.5

- PySpark 3.0.0

- matplotlib 3.2.1

- seaborn 0.10.1

- sklearn 0.22.1.post1

With that, let's begin. Again, we will not provide commentary as detailed as in the scikit-learn example, so be sure to review the whole process in scikit-learn to get a good idea of how it will go. Additionally, we won't be validating the model in PySpark in this example.

Data Processing

Here are the `import` statements:

```python
import pyspark #
from pyspark.sql import SparkSession
from pyspark import SparkConf, SparkContext
from pyspark.sql.types import *
from pyspark.ml.feature import VectorAssembler
from pyspark.ml import Pipeline
from pyspark.ml.classification import LogisticRegression as
LogisticRegressionPySpark
import pyspark.sql.functions as F
import os
import seaborn as sns
import sklearn #
from sklearn.metrics import confusion_matrix
from sklearn.metrics import roc_auc_score

import matplotlib #
import matplotlib.pyplot as plt

os.environ["SPARK_LOCAL_IP"]='127.0.0.1'
spark = SparkSession.builder.master("local[*]").getOrCreate()
spark.sparkContext._conf.getAll()

print("pyspark: {}".format(pyspark.__version__))
print("matplotlib: {}".format(matplotlib.__version__))
print("seaborn: {}".format(sns.__version__))
print("sklearn: {}".format(sklearn.__version__))
```

The output should look something like in Figure 2-20.

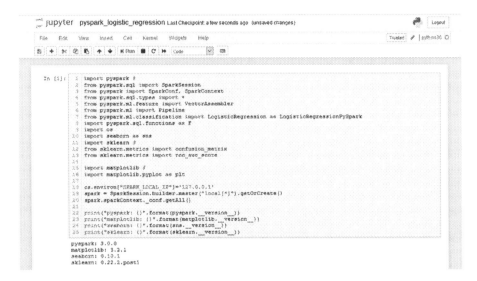

Figure 2-20. *Importing the necessary modules and printing their versions. Once again, modules imported solely for the purpose of displaying versions are marked with a # so you may remove them and the print statements if desired*

You will notice that there is some additional code relating to PySpark that you have had to define. With PySpark, you must define a Spark context and create a Spark session. What this really means is that you are creating a point of connection to the Spark engine, enabling the engine to run all of the code relating to Spark functionality.

Let's now load the data set. PySpark has its own functionality for creating data frames, so you won't be using pandas. Execute the following:

```
data_path = 'data/creditcard.csv'

df = spark.read.csv(data_path, header = True, inferSchema = True)
labelColumn = "Class"
columns = df.columns
numericCols = columns
numericCols.remove(labelColumn)
print(numericCols)
```

You should see something like Figure 2-21.

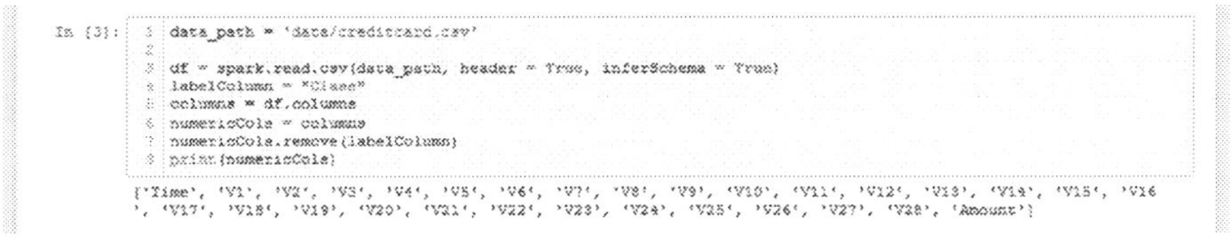

Figure 2-21. *Reading the credit card data set in PySpark and removing the Class column from the list of columns. This is done because you don't want the Class column to be included in the feature vector, as you will see in Figure 2-22*

Printing the columns is just to ensure that the label column has been removed successfully.

You can look at the data frame now just to ensure that it has been loaded properly. You will have to use built-in functionality to convert to a pandas data frame, because Spark data frames are not very clean to look at.

Look at the following two cells and their outputs:

```
df.show(2)
```

Refer to Figure 2-22.

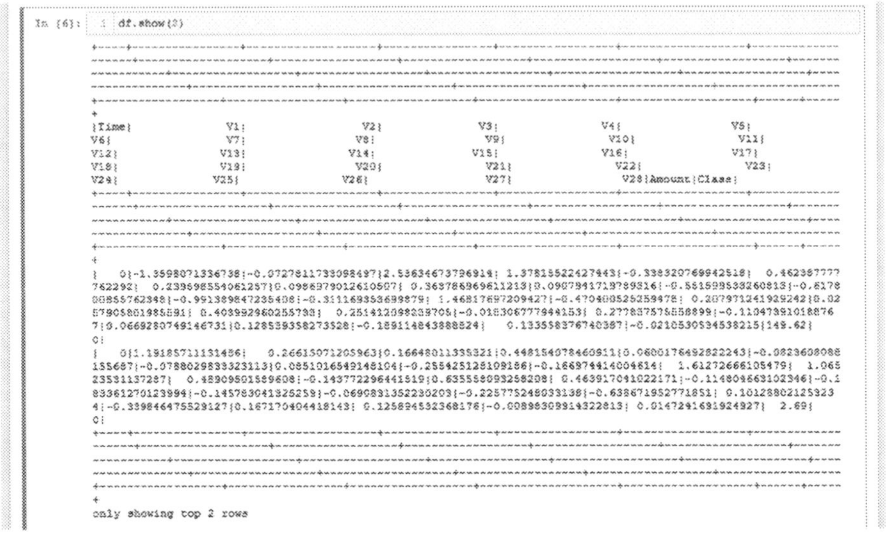

Figure 2-22. *The output of the Spark data frame. Since there are so many columns in the data frame, the output is very messy and very difficult to read. Fortunately, there is built-in functionality to convert PySpark data frames into pandas data frames, making it much easier to view the rows in the Spark data frame*

Now compare this to the following:

```
df.toPandas().head()
```

Refer to Figure 2-23.

```
In [9]:    1  df.toPandas().head()
```

Out[9]:

	Time	V1	V2	V3	V4	V5	V6	V7	V8	V9	...	V21	V22	V23	V24	
0	0	-1.359807	-0.072781	2.536347	1.378155	-0.338321	0.462388	0.239599	0.098698	0.363787	...	-0.018307	0.277838	-0.110474	0.066928	0.128
1	0	1.191857	0.266151	0.166480	0.448154	0.060018	-0.082361	-0.078803	0.085102	-0.255425	...	-0.225775	-0.638672	0.101288	-0.339846	0.167
2	1	-1.358354	-1.340163	1.773209	0.379780	-0.503198	1.800499	0.791461	0.247676	-1.514654	...	0.247998	0.771679	0.909412	-0.689281	-0.327
3	1	-0.966272	-0.185226	1.792993	-0.863291	-0.010309	1.247203	0.237609	0.377436	-1.387024	...	-0.108300	0.005274	-0.190321	-1.175575	0.647
4	2	-1.158233	0.877737	1.548718	0.403034	-0.407193	0.095921	0.592941	-0.270533	0.817739	...	-0.009431	0.798278	-0.137458	0.141267	-0.206

5 rows × 31 columns

Figure 2-23. *Using PySpark's built-in functionality to convert the spark data frame into a pandas data frame for easier viewing. As seen in Figure 2-22, it is extremely hard to read the direct output of a Spark data frame*

So whenever you want to check a Spark data frame, make sure to convert it to pandas if it has a lot of columns.

The data processing procedure for PySpark is slightly different than in pandas. To train the model, you must pass in a vector called features. Take a look at the following code:

```
stages = []
assemblerInputs = numericCols
assembler = VectorAssembler(inputCols=assemblerInputs,
outputCol="features")
stages += [assembler]

dfFeatures = df.select(F.col(labelColumn).alias('label'),
*numericCols )
```

This defines the inputs to the assembler so that it knows what columns to transform into the features vector.

From here, let's add to the cell above and create the normal and anomaly data splits as with the scikit-learn example.

```
normal = dfFeatures.filter("Class == 0").
sample(withReplacement=False, fraction=0.5, seed=2020)
anomaly = dfFeatures.filter("Class == 1")
```

```
normal_train, normal_test = normal.randomSplit([0.8, 0.2],
seed = 2020)
anomaly_train, anomaly_test = anomaly.randomSplit([0.8, 0.2],
seed = 2020)
```

The cell should look like Figure 2-24.

```
In [5]:   1
          2  stages = []
          3  assemblerInputs = numericCols
          4  assembler = VectorAssembler(inputCols=assemblerInputs, outputCol="features")
          5  stages += [assembler]
          6
          7  dfFeatures = df.select(F.col(labelColumn).alias('label'), *numericCols )
          8
          9  normal = dfFeatures.filter("Class == 0").sample(withReplacement=False, fraction=0.5, seed=2020)
         10  anomaly = dfFeatures.filter("Class == 1")
         11
         12  normal_train, normal_test = normal.randomSplit([0.8, 0.2], seed = 2020)
         13  anomaly_train, anomaly_test = anomaly.randomSplit([0.8, 0.2], seed = 2020)
```

Figure 2-24. *Constructing the VectorAssembler that will be used later to create a feature vector from the input data. You also create a normal and anomaly data split similar to how it was done in scikit-learn, and split it in a similar fashion into training and testing subsets*

Just like in the scikit-learn example, you combine the respective normal and anomaly splits to form your training and testing sets. This time, however, you won't have a validation set, so you are looking at an 80-20 split between the training and testing data.

```
train = normal_train.union(anomaly_train)
test = normal_test.union(anomaly_test)
```

Refer to Figure 2-25 to see the cell.

```
In [6]:   1  train = normal_train.union(anomaly_train)
          2  test = normal_test.union(anomaly_test)
```

Figure 2-25. *Creating the training and testing sets in a similar manner to how you did it in scikit-learn, but with PySpark's functionality*

Let's finish the rest of the pipeline and create the feature vector:

```
pipeline = Pipeline(stages = stages)
pipelineModel = pipeline.fit(dfFeatures)
train = pipelineModel.transform(train)
test = pipelineModel.transform(test)
selectedCols = ['label', 'features'] + numericCols
train = train.select(selectedCols)
test = test.select(selectedCols)

print("Training Dataset Count: ", train.count())
print("Test Dataset Count: ", test.count())
```

Refer to Figure 2-26 to see the output.

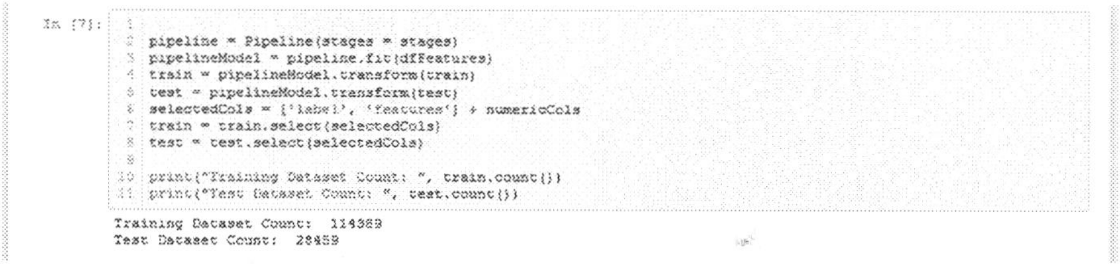

Figure 2-26. *Using a pipeline to create a feature vector from the data frame. This feature vector is what the logistic regression model will train on*

Model Training

You can now define and train the model:

```
lr = LogisticRegressionPySpark(featuresCol = 'features',
labelCol = 'label', maxIter=10)

lrModel = lr.fit(train)

trainingSummary = lrModel.summary
pyspark_auc_score = trainingSummary.areaUnderROC
```

73

Refer to Figure 2-27 to see the above code in a cell.

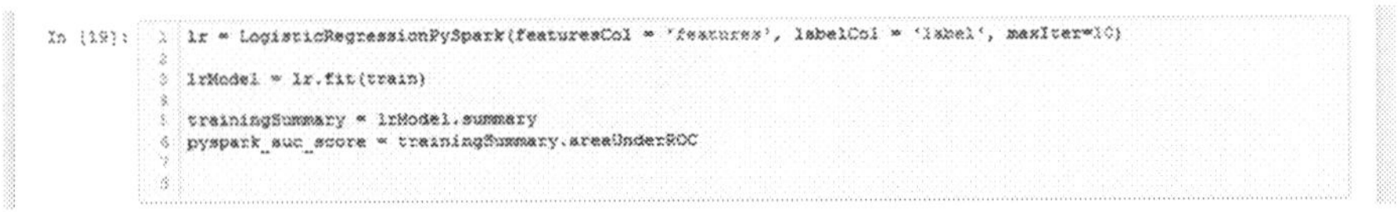

```
In [19]:   1  lr = LogisticRegressionPySpark(featuresCol = 'features', labelCol = 'label', maxIter=10)
           2
           3  lrModel = lr.fit(train)
           4
           5  trainingSummary = lrModel.summary
           6  pyspark_auc_score = trainingSummary.areaUnderROC
           7
           8
```

Figure 2-27. *Defining the PySpark logistic regression model, training it, and finding the AUC score using the built-in function of the model*

Model Evaluation

Once the model has finished training, run the evaluation code:

```
predictions = lrModel.transform(test)

y_true = predictions.select(['label']).collect()
y_pred = predictions.select(['prediction']).collect()

evaluations = lrModel.evaluate(test)
accuracy = evaluations.accuracy
```

Add the following code as well to display the metrics:

```
print(f"AUC Score: {roc_auc_score(y_pred, y_true):.3%}")
print(f"PySpark AUC Score: {pyspark_auc_score:.3%}")
print(f"Accuracy Score: {accuracy:.3%}")
```

Refer to Figure 2-28 to see the output.

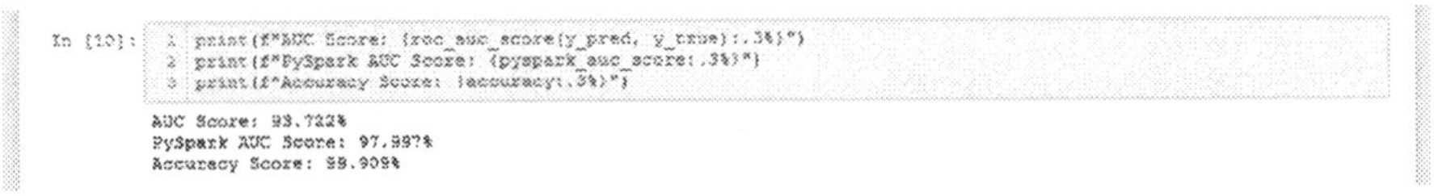

```
In [10]:   1  print(f"AUC Score: {roc_auc_score(y_pred, y_true):.3%}")
           2  print(f"PySpark AUC Score: {pyspark_auc_score:.3%}")
           3  print(f"Accuracy Score: {accuracy:.3%}")

AUC Score: 93.722%
PySpark AUC Score: 97.997%
Accuracy Score: 99.909%
```

Figure 2-28. *The output metrics. The AUC score is calculated using scikit-learn's scoring algorithm, while the PySpark AUC score metric comes from the training summary of the PySpark model. Finally, the accuracy score is also outputted*

You can see that the AUC score and the accuracy are quite high, so let's examine the graphs.

First, let's look at the ROC curve:

```
pyspark_roc = trainingSummary.roc.toPandas()
plt.xlabel('False Positive Rate')
plt.ylabel('True Positive Rate')
plt.title('PySpark ROC Curve')
plt.plot(pyspark_roc['FPR'],pyspark_roc['TPR'])
```

To see the graph, refer to Figure 2-29.

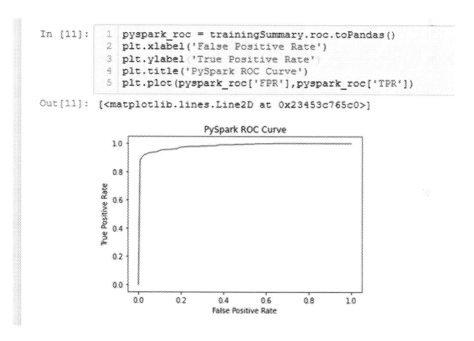

Figure 2-29. *The ROC curve for the PySpark logistic regression model you just trained. A perfect ROC curve would have the true positive rate starting at 1.0, where it continues right to a false positive rate value of 1.0. This curve is quite close to that, hence why its area (AUC) is said to be around 0.97997 by PySpark, keeping in mind a perfect AUC score is 1.00*

The curve looks quite optimal. Let's now look at the confusion matrix to get a detailed idea of how the model performs:

```
conf_matrix = confusion_matrix(y_true, y_pred)
ax = sns.heatmap(conf_matrix, annot=True, fmt='g')
ax.invert_xaxis()
ax.invert_yaxis()
plt.ylabel('Actual')
plt.xlabel('Predicted')
```

Refer to Figure 2-30 to view the confusion matrix plot.

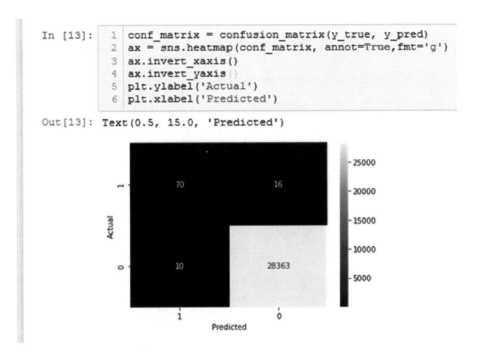

Figure 2-30. *The plotted confusion matrix of the PySpark logistic regression model you just trained. The accuracy of correctly labeled points for the normal data is very high and is decent for the anomalous data*

From this, you have a much more detailed account of how the model performed. Looking at just the anomalies, you see that the model has a 81.4% accuracy (70/(70+16)) in predicting anomalies. This is better than the model you trained in scikit-learn, though you haven't tuned the hyperparameter to attain maximum performance.

PySpark does have an option to weight your data, but this is done on a sample-by-sample basis. What this means is that instead of passing in a weight dictionary for each class, you have to create a column in the data frame with each anomaly being weighted a certain amount and each normal point being weighted as 1, for example. By default, everything is weighted as 1, so that means the PySpark model may have a greater potential in performance than the scikit-learn model.

Moving on to the normal points, you see a really good accuracy of 99.96% (28363/(28363+10)), so it is able to identify normal points very well.

Summary

With the insights you gained from data analysis, you processed the data into training, testing, and validation sets in scikit-learn and PySpark (you only did a train-test split in PySpark, but you could have split the training data into training and validation sets just like in scikit-learn). From there, you constructed logistic regression models in each framework and trained and evaluated on them. You looked at accuracy and AUC scores as metrics and looked at the ROC curve and confusion matrix to get a better idea of how the model performed. For the scikit-learn model, you performed k-fold cross-validation to help tune the hyperparameter. In the next chapter, you will keep your experiences with data analysis and model creation in mind as you learn about MLOps and how you can operationalize your models.

CHAPTER 3

What Is MLOps?

In this chapter, we will cover the concepts behind the term "MLOps" and go over what it is, why it's useful, and how it's implemented.

Introduction

Creating machine learning solutions to various problems can be quite the arduous task. Let's imagine ourselves in the shoes of a team that is attempting to solve a problem with machine learning. You may be familiar with this process if you read Chapter 1, but we will recap the entire process once again to establish the context. You may skip past this section if you are already familiar with this. The entire process may look somewhat like the following:

- **Collect and process raw data:** Raw data is rarely in a format that is easy to train a model on. Usually, it requires processing to remove aberrant data points such as null values and faulty data values. Other times, you might have to process the raw data to extract only the information you need among all of the noise.

© Sridhar Alla, Suman Kalyan Adari 2021
S. Alla and S. K. Adari, *Beginning MLOps with MLFlow*,
https://doi.org/10.1007/978-1-4842-6549-9_3

- **Analyze the data:** This step involves looking at the data points and understanding their characteristics. How is it structured? What does the distribution of the data points look like? Are there any identifiable trends or biases in the data? This step is crucial because it dictates how you are going to approach the problem. If you already have a trained model you are looking to update, it also tells you if there are any new trends in the data that your model should be updated to consider. If you identified any "useless features" that don't really influence the output, you might drop them and train a new model to improve training speed while possibly boosting performance.

- **Process the data for training:** In this step, you could be scaling the data to a more appropriate range and perhaps removing any outliers and/or anomalies that could interfere with model performance. Furthermore, you could also be applying feature engineering to create new features from existing data points and perhaps give your model more or a better context during training. This is also where you create training and testing data sets, though optimal practice is to make training, testing, and validation data sets.

- **Construct, train, and test the model:** In this step, you are creating the model, setting hyperparameters, and training the model. In the case of deep learning, you can also select a subsection of the training set to be a data validation set. The purpose of this set is to have the model be evaluated on it at the end of every epoch or full forward pass of the data through the model. By comparing the model performance on data it's seen many times over during training versus data it hasn't seen at all (or rather, data that has no effect on weight adjustment), you can see if the model is truly learning to generalize or if it's just overfitting.

- **Overfitting** is when a model performs significantly better on a training set compared to data that it has never seen before. As just discussed, one way to give an early indication of overfitting is to set aside a portion of the training set as validation data during the training phase. This can give you an early indication of overfitting without having to find out after the training process has finished, which can take anywhere from minutes to days depending on the depth of the model and the equipment used. And so, it follows that overfitting can also be observed when the model is evaluated on the testing data or validation data, and discrepancies in model performance can be observed between these sets and the training set.

- This phenomenon of overfitting could partially result from the model not receiving enough data points during training to reflect the variety it is expected to see, so fixing the training set by introducing more variety or even increasing the number of data points can help. Additionally, including methods such as regularization or dropout into the model's architecture can also help combat overfitting in the case of deep learning models.

- An important thing to discuss is the purpose of the **testing** and **validation** sets. Testing sets are reserved for evaluating a model's performance on data it's never seen before.

- Validation sets are reserved for helping select models, select model architecture, tune hyperparameters, or simply to give an indication of model performance on data it's never seen during the training process.

- An example of validation is k-fold cross-validation, where it generates k random partitions of test-train data from validation data and can be used to train/ evaluate the model on all of them to give an idea of the best performance it can attain with various hyperparameter settings. Of course, we can also use k-fold cross validation to perform the other functions that validation helps with. You looked at an example using this method of validation in Chapter 2, when you used it to help tune the weighting of anomalies.

- Coupling this technique with a script that has a set of hyperparameters can result in an optimal model with proper hyperparameters. From there, the model can be retrained and evaluated again on the test set to get a final performance benchmark.

- The specific order this is done in can differ, though. For example, trained models can also be evaluated first and then validated, compared to the other way around. This is because the training process is likely to be repeated with altered hyperparameters anyway after the evaluation stage reflects some form of performance discrepancy or if validation data during the training process reveals that possible overfitting is occurring. Either way, it really depends, but good practice is to at least incorporate both testing and validation data to best tune the model.

- **Validate and tune the model:** As previously discussed, the validation set can be another "testing" set that the model has never seen before, and can be used in any of the several ways described earlier and in Chapter 1. Once your model has reached an acceptable level of performance on the validation set and is retrained and evaluated again, you can look at deploying the model.

- **Deploy and monitor the model:** In this step, the model has finally left the hands of the machine learning/data science team. It is now the job of engineering and operational teams to integrate this model into the application and put it into service. Operational teams are in charge of constantly monitoring the performance of the model, with dips in performance possibly indicating that this entire process may need to be repeated to update the model to understand new trends. Operational teams are also responsible for reporting any bugs and unexpected model predictions to the data science team, feedback that also contributes to the start of this whole cycle as the model needs to be fixed.

Hopefully, it's clear just how work-intensive the entire process can get, especially since it will most likely need to be repeated multiple times. While it is possibly easier the second time around since you're only updating the model on new data patterns and trends, it is still a problem that can take up hours of manual labor that can be better spent elsewhere. After all, maintenance of applications in the software development process is usually where most of the money and resources go, not the initial construction and release of the application. The same can apply to machine learning models, worsening the overall maintenance costs because the costs for deployed machine learning models are added on top of the costs for the software application utilizing the services of the models.

Imagine if you could simply automate this entire process away, allowing you to take full advantage of high-performance machine learning models without all of that hassle. This is where MLOps comes in, something that can be thought of as the intersection between machine learning and DevOps practices. **DevOps**, or developmental operations, refers to a set of practices that combines the work processes of software developers with those of operational teams to create a common set of practices that functions as a hybrid of the two roles. As a result, the developmental cycle of software is expedited, and continuous delivery of software products is ensured. Total costs also go down because maintenance costs are reduced as a result of the increase in efficiency of the workflow in maintaining the software applications. Refer to Figure 3-1 to see a graph representing the DevOps workflow.

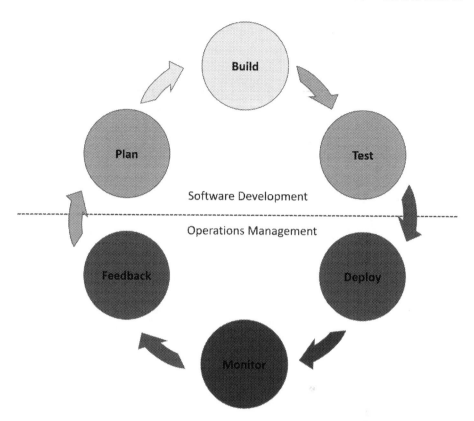

Figure 3-1. *A graph depicting the workflow in a DevOps environment. Software development teams typically adopt the Agile methodology of software development, which is summarized above through the planning, building, and testing stages. Operational teams are in charge of deploying, maintaining, and collecting feedback in the form of bugs and user feedback and relaying this information to the development teams. From there, the development team enters the maintenance phase of the application, where they plan, build, test, and push the next patch/update for the application. Furthermore, automating the process of testing and deploying allows for continuous integration and delivery of software products, something we will expand upon later in this chapter*

Similarly, MLOps adopts DevOps principles and applies them to machine learning models in place of software, uniting the developmental cycles followed by data scientists and machine learning engineers with that of operational teams to help ensure continuous delivery of high-performance machine learning models. The process of model development in what's called the **experimental stage**, something we will look at in detail later in the chapter, can lead to impressive performances and can seem like very promising solutions. However, the reality is more that most models simply never make it past this experimental stage, since deploying them is a massive undertaking on its own. Unfortunately, maintaining models once deployed also drains resources, as every new update requires reintegration into the application. This means that even if the model is deployed, all teams have their work cut out for them. For these reasons, most models simply never make it past the prototype phase.

Until the emergence of MLOps principles, deploying solutions created using the latest in machine learning technology served as a significant challenge to businesses due to the amount of resources that would be required. **This is why MLOps is so crucial**. It makes it significantly easier to deploy and maintain your machine learning solutions by automating most of the hard parts for you, massively expediting the development and maintenance processes. With a fully automated setup, teams can keep up with the latest in machine learning technology and deploy new models quickly. Services can maintain their high level of performance and perhaps even improve on this front as teams can deploy newer, more promising model architectures.

Now that you have a better idea of what MLOps is about and why it is so important, let's jump into the details and look at how an ideal MLOps implementation is set up.

MLOps Setups

Before we look at any specific MLOps setups, let's first establish three different setups representing the various stages of automation: **manual implementation**, **continuous model delivery**, and **continuous integration/continuous delivery of pipelines**.

Manual implementation refers to a setup where there are no MLOps principles applied and everything is manually implemented. The steps discussed above in the creation of a machine learning model are all manually performed. Software engineering teams must manually integrate the models into the application, and operational teams must help ensure all functionality is preserved along with collecting data and performance metrics of the model.

Continuous model delivery is a good middle ground between a manual setup and a fully automated one. Here, we see the emergence of **pipelines** to allow for automation of the machine learning side of the process. Note that we will mention this term quite often in the sections below. If you'd like to get a better idea about what a pipeline is, refer to the section titled "Pipelines and Automation" further down in this chapter. For now, a **pipeline** is an infrastructure that contains a sequence of components manipulating information as it passes through the pipeline. The function of the pipeline can slightly differ within the setups, so be sure to refer to the graphs and explanations to get a better idea of how the pipeline in the example functions.

The main feature of this type of setup is that the deployed model has pipelines established to continuously train it on new data, even after deployment. Automation of the experimental stage, or the model development stage, also emerges along with modularization of code to allow for further automation in the subsequent steps. In this setup, **continuous delivery** refers to expedited development and deployment of new machine learning models. With the barriers to rapid deployment lifted (the tediousness of manual work in the experimental stage) by automation, models can now be created or updated at a much faster pace.

Continuous integration/continuous delivery of pipelines refers to a setup where pipelines in the experimental stage are thoroughly tested in an automated process to make sure all components work as intended. From there, pipelines are packaged and deployed, where deployment teams deploy the pipeline to a test environment, handle additional testing to ensure both compatibility and functionality, and then deploy it to the production environment. In this setup, pipelines can now be created and deployed at a quick pace, allowing for teams to continuously create new pipelines built around the latest in machine learning architectures without any of the resource barriers associated with manual testing and integration.

Manual Implementation

Now that we've established three variations of setups, let's look at the first of the three deployment setups of machine learning models, which has no MLOps principles integrated.

In this case, there is a team of data scientists and machine learning engineers, who will now be referred to as the "model development team," manually performing data analysis and building, training, testing, and validating their models. Once their model has been finalized, they must create a model class and push this to a code repository. Software engineers extract this model class and integrate it into an existing application or system, and operational teams are in charge of monitoring the application, maintaining functionality, and providing feedback to both the software and model development teams.

Everything here is manual, meaning any new trends in the data lead to the model development team having to update the model and repeat the entire process again. This is quite likely to happen considering the high volume of users interacting with your model every day. Combined with performance metrics and user data collection, the information will reveal a lot of aspects about your model as well as the user base the model is servicing. Chances are high that you will have to update it to maintain its performance on the new data. This is something to keep in mind as you follow through with the process on the graph.

Refer to Figure 3-2 for a graphical representation of the setup.

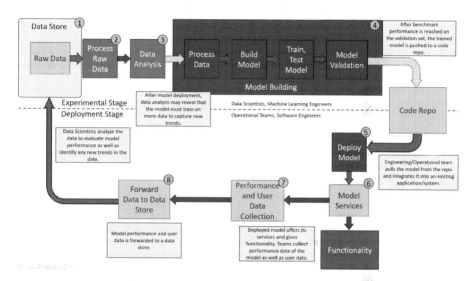

Figure 3-2. *Graph depicting a possible deployment setup of a machine learning model without MLOps principles. The arrows with a dotted border mean that progression to the next step depends upon a condition in the current step. For example, in the model validation step, machine learning engineers must ensure that the model meets a minimum benchmark in performance before pushing a model class to the repository*

Let's go through this step by step. We can split the flow into roughly two parts: the **experimental stage**, which involves the machine learning side of the entire workflow, and the **deployment stage,** which handles integration of the model into the application and maintaining operations.

Experimental Stage:

1. **Data store:** The data store refers to wherever data relevant to data analysis and model development is stored. An example of a data store could be using Hadoop to store large volumes of data, which can be used by multiple model development teams. In this example, data scientists can pull raw data from this data store to start performing experiments and conducting data analysis.

2. **Process raw data:** As previously mentioned, raw data must be processed in order to collect the relevant information. From there, it must also be purged of faults and corrupted data. When a company collects massive volumes of data every day, some of it is bound to be corrupted or faulty in some way eventually, and it's important to get rid of these points because they can harm the data analysis and model development processes. For example, one null value entry can completely destroy the training process of a neural network used for a regression (value prediction) task.

3. **Data analysis:** This step involves analyzing all aspects of the data. The general gist of it was discussed earlier, but in the context of updating the model, data scientists want to see if there are any new trends or variety in data that they think the model should be updated on. Since the initial training process can be thought of as a small representation of the real-world setting, there is a fair chance that the model will need to be updated

soon after the initial deployment. This does depend on how many characteristics of the true user base the original training set captured however, but user bases change over time, and so must the models. By "user base," we refer to the actual customers using the prediction services of the model.

4. **Model building stage:** This stage is more or less the same as what we discussed earlier. The second time around, when updating the model, it could turn out that slight adjustments to the model layers may be needed. In some of the worst cases, the current model architecture being used cannot achieve a high enough performance even with new data or architectural tweaks. An entirely new model may have to be built, trained, and validated. If there are no such issues, then the model would just be further trained, tested, validated, and pushed to the code repository upon meeting some performance criteria.

- An important thing to note about this experimental stage is that it is quite popular for experiments to be conducted using Jupyter notebook. When model development teams reach a target level of performance, they must work on building a workable model that can be called by other code. For example, this can be done by creating a model class with various functions that provide functionality such as `load_weights`, `predict`, and perhaps even `evaluate` to allow for easier gathering of performance metrics. Since the true label can't be known in real-time settings, evaluation metrics can simply be something like a root-mean-squared error.

Deployment Stage:

5. **Model deployment:** In this case, this step is where
   software engineers must manually integrate
   the model into the system/application they are
   developing. Whenever the model development
   team finishes with their experiments, builds
   a workable model, and pushes it to the code
   repository, the engineering team must manually
   integrate it again. Although the process may not be
   that bad the second time around, there is still the
   issue of fixing any potential bugs that may arise from
   the new model. Additionally, engineering teams
   must also handle testing of not only the model once
   it is integrated into the application, but also of the
   rest of the application.

6. **Model services:** This step is where the model is
   finally deployed and is interacting with the user
   base in real time. This is also where the operational
   team steps in to help maintain the functionality of
   the software. For example, if there are any issues
   with some aspect of the model functionality, the
   operational team must record the bug and forward it
   to the model development team.

7. **Data collection:** The operational team can also
   collect raw data and performance metrics. This
   data is crucial for the company to operate since
   that is how it makes its decisions. For example, the
   company might want to know what service is most
   popular with the user base, or how well the machine
   learning models are performing so far. This job can

be performed by the application as well, storing all the relevant data in some specific data store related to the application.

8. **Data forwarded to data store:** This step is where the operational team sends the data to the data store. Because there could be massive volumes of data collected, it's fair to assume some degree of automation on behalf of the operational team on this end. Additionally, the application itself could also be in charge of forwarding data it collects to the relevant data store.

Reflection on the Setup

Right away, you can notice some problems that may arise from such an implementation. The first thing to realize is that the entire experimental stage is manual, meaning data scientists and machine learning engineers must repeat those steps every time. When models are constantly exposed to new data that is more than likely not captured in the original training set, models must frequently be retrained so that they are always up to date with current trends in user data. Unfortunately, when the entire process of analyzing new trends, training, testing, and validating data is manual, this may require significant resources over time, which may become unfeasible for a company without the resources to spare. Additionally, trends in data can change over time. For example, perhaps the age group with the largest number of users logging into the site is comprised of people in their early twenties. A year later, perhaps the dominant age group is now teenagers. What was normal back then isn't normal now, and this could lead to losses in ad revenues, for example, if that's the service (targeted advertising) the model in this case provides.

Another issue is that tools such as Jupyter notebook are very popular for prototyping and experimenting machine learning and deep learning models. Even if the experiments aren't carried out on notebooks, it's likely that work must be done in order to push the model to the source repo. For example, constructing a model class with some important functions such as `load_weights`, `predict`, and `evaluate` would be ideal for a model class. Some external code may call upon `load_weights()` to set the model weights from different training instances (so if the model has been further trained and updated, simply call this function to get the new model). The function `predict()` would then be called to make predictions based on some input data and provide the services the application requires, and the function `evaluate()` would be useful in keeping performance metrics. Live data will almost never have truth labels on it (unless the user provides instant feedback, like Google's captchas where you select the correct images), so a score metric like a root-mean-squared error can be useful when keeping track of performance.

Once the model class is completed and pushed, software engineering teams must integrate the model class into the overall application/system. This could prove difficult the first time around, but once the integration has been completed, updates to the model can be as simple as loading new weights. Unfortunately, model architectures are likely to change, so the software teams must reintegrate new model classes into the application.

Furthermore, deep learning is a complicated and rapidly evolving field. Models that were cutting-edge several years ago can be far surpassed by the current state-of-the-art models, so it's important to keep updating your model architectures and to make full use of the new developments in the field. This means teams must continuously repeat the model-building process in order to keep up with developments in the field.

Hopefully it is more clear that this implementation is quite flawed in how much work is required to not only create and deploy the model in the first place, but also to continuously maintain it and keep it up to par.

Alright, so how would we go about improving it? Where does this MLOps come into play? To answer these questions, let's look at the second setup of the three defined earlier.

Continuous Model Delivery

This setup contains **pipelines** for automatic training of the deployed model as well as for speeding up the experimental process. Refer to Figure 3-3 for a graphical representation of this setup.

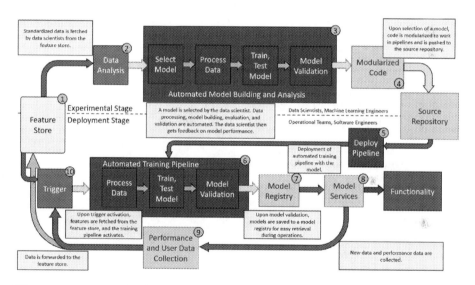

Figure 3-3. *Graph depicting a possible deployment setup of a machine learning model with automation via pipelines*

This is a lot to take in at once, so let's break it down and follow it according to the numbers on the graph.

1. **Feature store:** This is a data storage bin that takes the place of the data store in the previous example. The reason for this is that all data can now be standardized to a common definition that all

processes can use in this instance. For example, the processes in the experimental stage will be using the same input data as the deployed training pipeline because all of the data is held to the same definition. What is meant by **common definition** is that raw data is cleansed and processed in a procedural way that applies to all relevant raw data. These processed features are then held in the feature store for pipelines to draw from, and it is ensured that every pipeline uses features processed according to this standard. This way, any perceived differences in trends between two different pipelines won't be attributed to deviances in processing procedures.

Presume for an instance that you are trying to provide an object detection service that detects and identifies various animals in a national park. All video feed from the trail cameras (a video can be thought of as a sequence of frames) can be stored as raw data, but it can be possible that different trail cameras have different resolutions. Instead of repeating the same data processing procedure, you can simply apply the same procedure (normalizing, scaling, and batching the frames, for example) to the raw videos and store the features that you know all pipelines will use.

2. **Data analysis:** In this step, data analysis is still performed to give data scientists and machine learning engineers an idea of what the data looks like, how it's distributed, and so on, just like in the manual setup. Similarly, this step can determine whether or not to proceed with construction of a new model or just update the current model.

3. **Automated model building and analysis:** In this
    step, data scientists and machine learning engineers
    can select a model, set any specific hyperparameters,
    and let the pipeline automate the entire process.
    The pipeline will automatically process the data
    according to the specifications of this model (take
    the case where the features are 331x331x3 images
    but this particular model only accepts images that
    are 224x224x3), build the model, train it, evaluate
    it, and validate it. During validation, the pipeline
    may automatically tune the hyperparameters
    as well optimize performance. It is possible that
    manual intervention may be required in some
    cases (debugging, for example, when the model is
    particularly large and complex, or if the model has a
    novel architecture), but automation should otherwise
    take care of producing an optimal model. Once this
    occurs, modularized code is automatically created so
    that this pipeline can be easily deployed.

    Everything in this stage is set up so that the
    experimental stage goes very smoothly, requiring
    only that the model is built. Depending on the level of
    automation implemented, perhaps all that is required
    is that the model architecture is selected with some
    hyperparameters specified, and the automation takes
    care of the rest. Either way, the development process
    in the experimental stage is sped up massively. With
    this stage going faster, more experiments can be
    performed too, leading to possible boosts in overall
    efficiency as productivity is increased and optimal
    solutions can be found quicker.

4. **Modularized code:** The experimental stage is set up so that the pipeline and its components are modularized. In this specific context, the data scientist/machine learning engineer defines and builds some model, and the data is standardized to some definition. Basically, the pipeline should be able to accept any constructed model and perform the corresponding steps given some data without hardcoding anything. (Meaning there isn't any code that will only work for a specific model and specific data. The code works with generalized cases of models and data.)

   This is **modularization**, when the whole system is divided into individual components that each have their own function, and these components can be switched out depending on variable inputs. Thanks to the modularized code, when the pipeline is deployed, it will be able to accept any new feature data as needed in order to update the deployed model. Furthermore, this structure also lets it swap out models as needed, so there's no need to construct the entire pipeline for every new model architecture.

   Think of it this way: the pipeline is a puzzle piece, and the models along with their feature data are various puzzle pieces that can all fit within the pipeline. They all have their own "image" on the piece and the other sides can have variable shapes, but what is important is that they fit with the pipeline and can easily be swapped out for others.

5. **Deploy pipeline:** In this step, the pipeline is manually deployed and is retrieved from the source code. Thanks to its modularization, the pipeline setup is able to operate independently and automatically train the deployed model on any new data if needed, and the application is built around the code structure of the pipeline so all components will work with each other correspondingly. The engineering team has to build parts of the application around the pipeline and its modularized components the first time around, but after that, the pipelines should work seamlessly with the applications so as long as the structure remains the same. Models are simply swapped, unlike before when the model had to be manually integrated into the application. This time, the pipeline must be integrated into the application, and the models are simply swapped out.

   However, it is important to mention that pipeline structures can change depending on the model. The main takeaway here is that pipelines should be able to handle many more models before having to be restructured compared to the setup before where "swapping" models meant you only loaded updated weights. Now, if several architectures all have common training, testing, and validation code, they can all be used under the same pipeline.

6. **Automated training pipeline:** This pipeline contains the model that provides its services and is set up to automatically fetch new features upon activation of the trigger. The conditions for trigger

activation will be discussed in item 10. When the pipeline finishes updating a trained model, the model is saved to a model registry, a type of storage unit that holds trained models for ease of access.

7. **Model registry:** This is a storage unit that specifically holds model classes and/or weights. The purpose of this unit is to hold trained models for easy retrieval by an application, for example, and it is a good component to add to an automation setup. Without the model registry, the model classes and weights would just be saved to whatever source code repository is established, but this way, we make the process simpler by providing a centralized area of storage for these models. It also serves to bridge the gap between model development teams, software development teams, and operational teams since it is accessible by everyone, which is ultimately what we want in an ideal automation setup.

   This registry along with the automated training pipeline assures **continuous delivery of model services** since models can frequently be updated, pushed to this registry, and deployed without having to go through the entire experimental stage.

8. **Model services:** Here the application pulls the latest, best performing model from the model registry and makes use of its prediction services. This action then goes on to provide the desired functionality in the application.

9. **Performance and user data collection:** New data is collected as usual along with performance metrics related to the model. This data goes to the feature store, where the new data is processed and standardized so that it can be used in both the experimental stage and the deployment stage and there are no discrepancies between the data used by either stage. Performance data is stored so that data scientists can tell how the model is performing once deployed. Based on that data, important decisions such as whether or not to build a new model with a new architecture can be made.

10. **Training pipeline trigger:** This trigger, upon activation, initiates the automated training pipeline for the deployed model and allows for feature retrieval by the pipeline from the feature store. The trigger can have any of the following conditions, although it is not limited to them:

    - **Manual trigger:** Perhaps the model is to be trained only if the process is manually initiated. For example, data science teams can choose to start this process after reviewing performance and data and concluding that the deployed model needs to train on fresh batches of data.

    - **Scheduled training:** Perhaps the model is set to train on a specific schedule. This can be a certain time on the weekend, every night during hours of lowest traffic, every month, and so on.

- **Performance issues:** Perhaps performance data indicates that the model's performance has dipped below a certain benchmark. This can automatically activate the training process to attempt to get the performance back up to par. If this is not possible or is taking too many resources, data scientists and machine learning engineers can choose to build and deploy a new model.

- **Changes in data patterns:** Perhaps changes in the trends of the data have been noticed while creating the features in the feature store. Of course, the feature store isn't the only possible place that can analyze data and identify any new trends or changes in the data. There can be a separate process/program dedicated to this task, which can decide whether or not to activate the trigger.

  This would also be a good condition to begin the training process, since the new trends in the data are likely to lead to performance degradation. Instead of waiting for the performance hit to activate the trigger, the model can begin training on new data immediately upon sufficient detection of such changes in the data, allowing for the company to minimize any potential losses from such a scenario.

Reflection on the Setup

This implementation fixes many of the issues from the previous setup. Thanks to the integration of pipelines in the experimental stage, the previous problem of having the entire stage be composed of manual processes is no longer a concern. The pipeline automates the whole process of training, evaluating, and validating a model. The model development team now only needs to build the model and reuse any common training, evaluation, and validation procedures that are still applicable to this model. At the end of the model development pipeline, relevant model metrics are collected and displayed to the operator. These metrics can help the model development team to prototype quickly and arrive at optimal solutions even faster than they would have without the automation since they can run multiple pipelines on different models and compare all of them at once.

Automated model creation pipelines in the experimental stage allow for teams to respond faster to any significant changes in the data or any issues with the deployed model that need to be resolved. Unlike before, where the only model swapping was the result of loading updated weights for the same model, these pipelines are structured to allow for various models with different architectures as long as they all use the same training, evaluation, and validation procedures. Thanks to the modularized code, the pipeline can simply swap out model classes and their respective weights once deployed. The modularization allows for easier deployment of the pipeline and lets models be swapped out easily to allow for further training of any model during deployment. Should a model require special attention from the model development team, it can simply be trained further by the team and swapped back in once it is ready. Now teams can respond much more quickly by being able to swap models in and out in such a manner.

The pipelines also make it much easier for software engineering teams and operational teams to deploy the pipelines and models. Because everything is modularized, teams do not have to work on integrating new model classes into the application every time. Everyone benefits, and model development teams do not have to be as hesitant about implementing new architectures so as long as the new model still uses the same training, evaluation, and validation code as in the existing pipeline.

While this setup solves most of the issues that plagued the original setup, there are still some important problems that remain. Firstly, there are no mechanisms in place to test and debug the pipelines, so this must all be done manually before it is pushed to a source repository. This can become a problem when you're trying to push many iterations of pipelines, such as when you're building different models with architectures that differ in how they must be trained, tested, and validated. Perhaps the latest models are showing a vast improvement over the old state-of-the art, and your team wants to implement these new solutions as soon as possible. In situations like this, teams will frequently need to debug and test pipelines before pushing them to source code for deployment. In this case, there is still some automation left to be done to avoid manual work.

Pipelines are also manually deployed, so if the structure in the code changes, the engineering teams must rebuild parts of the application to work with the new pipeline and its modularized code. Modularization works smoothly when all components know what to expect from each other, but if the code of one of the components changes so that it isn't compatible anymore, either the application must be rebuilt to accommodate the new changes or the component must be rewritten to work with the original pipeline. Unfortunately, new model architectures may require that part of the pipeline itself be rewritten, so it is likely that the application itself must be worked on to accommodate the new pipeline.

Hopefully you begin to see the vast improvements that automation has made in this setup, but also the issues that remain to be solved. The automation has solved the issue of building and creating new models, but the problem of building and creating new pipelines still remains.

To find an answer to that problem, let's take a look at the last of the three setups defined earlier.

Continuous Integration/Continuous Delivery of Pipelines

In this setup, we will be introducing a system to thoroughly test pipeline components before they are packaged and ready to deploy. This will ensure **continuous integration of pipeline code** along with **continuous delivery of pipelines**, crucial elements of the automation process that the previous setup was missing. Refer to Figure 3-4 for a graphical representation of such a setup.

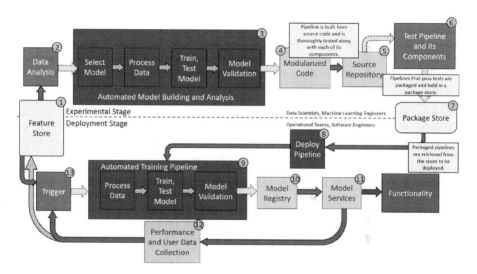

Figure 3-4. *Graph depicting added testing systems and a package store to the automation setup in Figure 3-2*

Though this is mostly the same setup, we will go through it again step by step with an emphasis on the newly introduced elements.

1. **Feature store:** The feature store contains standardized data processed into features. Features can be pulled by data scientists for offline data analysis. Upon activation of the trigger, features can also be sent to the automated training pipeline to further train the deployed model.

2. **Data analysis:** This step is performed by data scientists on features pulled from the feature store. The results from the analysis can determine whether or not to build a new model or adjust the architecture of an existing model and retrain it from there.

3. **Automated model building and analysis:** This step is performed by the model development team. Models can be built by the team and passed into the pipeline, assuming that they are compatible with the training, testing, and validation code, and the entire process is automatically conducted with a model analysis report generated at the end. In the case where the team wants to implement some of the latest machine learning architectures, models will have to be created from scratch with integration into pipelines in mind to maintain modularity. Parts of the pipeline code may have to change as well, which is acceptable because the new components of this setup can handle this automatically.

4. **Modularized code:** Once the model reaches a minimum level of performance in the validation step, the pipeline, its components, and the model are all ready to be modularized and stored in a source repository.

5. **Source repository:** The source repository holds all of the packaged pipeline and model code for different pipelines and different models. Teams can create multiples at once for different purposes and store them all here. In the old setup, pipelines and models would be pulled from here and manually

integrated and deployed by software engineering teams. In this setup, the modularized code must now be tested to make sure all of the components will work correctly.

6. **Testing:** This step is crucial in achieving **continuous integration**, or a result of automation where new components and elements are continuously designed, built, and deployed in the new environment.

    Pipelines and their components, including the model, must be thoroughly tested to ensure that all outputs are correct. Furthermore, the pipelines themselves must be tested so that they are guaranteed to work with the application and how it is designed. There shouldn't be bugs in the pipeline, for example, that would break its compatibility with the application. The application is programmed to expect a specific behavior from the pipeline, and the pipeline must behave correspondingly.

    If you are familiar with software development, the testing of pipeline components and the models is similar to the automated testing that developers write to check various parts of an application's functionality. A simple example is automated testing to ensure data of various types are successfully received by the server and are added to the correct databases.

With pipelines and machine learning models, some examples of testing include:

- Does the validation testing procedure lead to correct tuning of the hyperparameters?

- Does each pipeline component work correctly? Does it output the expected element? For example, after model evaluation, does it correctly begin the validation step? (Alternatively, if model evaluation goes after model validation, does the evaluation step correctly initiate?)

- Is the data processing performed correctly? Are there any issues with the data post-processing that would lead to poor model performance? Avoiding this outcome is for the best since it would waste resources having to fix the data processing component. If the business relies on rapid pipeline deployment, then avoiding this type of scenario is even more crucial.

- Does the data processing component correctly perform data scaling? Does it correctly perform feature engineering? Does it correctly transform images?

- Does the model analysis work correctly? You want to make sure that you're basing decisions on accurate data. If the model truly performs well but faults in the model analysis component of the pipeline lead the data scientist/machine learning engineer to believe the model isn't performing that well, then it could lead to issues where pipeline deployment is slowed down. Likewise, you don't want the model analysis to be displaying the wrong information, even if it mistakenly displays precision for accuracy.

The more thorough the automated testing, the better the guarantee that the pipeline will operate within the application without issues. (This doesn't necessarily include model performance as that has to do more with the model architecture, how the model is developed, and what it is capable of.)

Once the pipeline passes all the tests, it is then automatically packaged and sent to a package store. Continuous integration of pipelines is now achieved since teams can build modularized and tested pipelines much more quickly and have them ready for deployment.

7. **Package store:** The package store is a containment unit that holds various packaged pipelines. It is optional but included in this setup so that there is a centralized area where all teams can access packaged pipelines that are ready for deployment. Model development teams push to this package store, and software engineers and operational teams can retrieve a packaged pipeline and deploy it. In this way, it is similar to the model registry in that both help achieve **continuous delivery**. The package store helps achieve continuous delivery of pipelines just as the model registry helps achieve continuous delivery of models and model services.

Thanks to automated testing providing continuous integration of pipelines and continuous delivery of pipelines via the package store, pipelines can also be deployed rapidly by operational teams and software engineers. With this, businesses can easily keep up with the latest trends and advances in machine learning architectures, allowing for better and better performance and more involved services.

8. **Deploy pipeline:** Pipelines can be retrieved from the package store and deployed in this step. Software engineering and operational teams must ensure that the pipeline will integrate without incident into the application. Because of that, there can be more testing on the part of software engineering teams to ensure proper integration of the pipeline. For example, one test can be to ensure the dependencies of the pipeline are considered in the application (if, for example, TensorFlow has updated and contains new functionality the pipeline now uses, the application should update its version of TensorFlow as well).

   Teams usually want to deploy the pipelines into a test environment where it will be subjected to further automated testing to ensure full compatibility with the application. This can be done automatically, where the pipelines go from the package store into the test environment, or manually, where teams decide to deploy the pipeline into the test environment. After the pipeline passes all the tests, teams can choose to manually deploy the pipeline into the production environment or have it automatically done.

Either way, pipeline creation and deployment is a much faster process now especially since teams do not have to manually test the pipelines and they do not have to build or modify the application to work with the pipeline every time.

9. **Automated training pipeline:** The automated training pipeline, once deployed, exists to further train models upon activation of the trigger. This helps keep models as up to date as possible on new trends in data and maintain high performance for longer. Upon validation of the model, models are sent to the model registry where they are held until they are needed for services.

10. **Model registry:** The model registry holds trained models until they are needed for their services. Once again, continuous delivery of model services is achieved as the automated training pipeline continuously provides the model registry with high-performance machine learning models to be used to perform various services.

11. **Model services:** The best models are pulled from the model registry to perform various services for the application.

12. **Performance and user data collection:** Model performance data and user data is collected to be sent to model development teams and the feature store, respectively. Teams can use the model performance metrics along with the results from the data analysis to help decide their next course of action.

13. **Training pipeline trigger:** This step involves some condition being met (refer to the previous setup, *continuous model delivery*) to initiate the training process of the deployed pipeline and feed it with new feature data pulled from the feature store.

Reflection on the Setup

The main issue of the previous setup that this one fixes is that of pipeline deployment. Previously, pipelines had to be manually tested by machine learning teams and operational teams to ensure that the pipeline and its components worked, and that the pipeline and its components were compatible with the application. However, in this setup, testing is automated, allowing for teams to much more easily build and deploy pipelines than before. The biggest advantage to this is that businesses can now keep up with significant changes in the data requiring the creation of new models and new pipelines, and can also capitalize on the latest machine learning trends and architectures all thanks to rapid pipeline creation and deployment combined with continuous delivery of model services from the previous setup.

The important thing to understand from all these examples is that automation is the way to go. Machine learning technology has progressed incredibly far within the last decade alone, but finally, the infrastructure to allow you to capitalize on these advancements is catching up.

Hopefully, after seeing the three possible MLOps setups, you understand more about MLOps and how implementations of MLOps principles might look. You might have noticed that pipelines have been mentioned quite often throughout the descriptions of the setups, and you might be wondering, "What are pipelines, and why are they so crucial for automation?"

To answer that question, let's take a look at what a "pipeline" really is.

Pipelines and Automation

Pipelines are an important part of automation setups employing DevOps principles. One way to think about a **pipeline** is that it is a specific, often sequential procedure that dictates the flow of information as it passes through the pipeline. To see an example of a testing pipeline in a software development setting, refer to Figure 3-5.

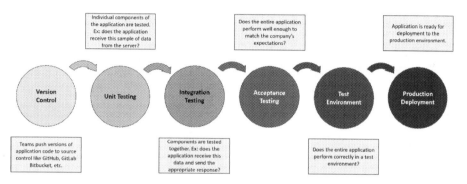

Figure 3-5. *A testing pipeline in a software development setting. The pipeline for testing packaged model pipelines in the optimal setup above is similar in that individual components must be tested, components must be tested in groups, and in the case where pipelines are deployed to a test environment first where further tests are performed before they are deployed to the production environment*

In the MLOps setups above, you've seen pipelines for automating the process of training a deployed model and for building, testing, and packing pipelines as well as for testing integration of packaged pipelines before deploying them to the production environment.

So, what does all that really mean? To get a better idea of what exactly goes on in a pipeline, let's follow the flow of data through a pipeline in the experimental stage. Even if you understand how pipelines work, it may be worth following the example anyway as we now look at this pipeline through the context of using MLOps APIs.

Journey Through a Pipeline

We will be looking at the model development pipeline in the experimental stage. Before we begin, it is important to mention that we will be referencing API calls in this pipeline. This is because some APIs can be called while executing scripts or even Jupyter cells at key points in the model's development, giving MLOps monitoring software information on model training, model evaluation, and model validation. At the end of the pipeline, the MLOps software would also ready the model for deployment via functionality provided by the API.

You will read more about this API in the next chapter, Chapter 4, but for now, you may assume that the API will take care of automation as you follow along through the example.

Model Selection

As seen in Figure 3-4, the experimental pipeline begins with the selection of a model. This is up to the operator, who must now choose and build a model. Some APIs allow you to call their functionality while building the model to connect with MLOps software as the rest of the process goes on. This software then keeps track of all relevant metrics related to the model's development along with the model itself in order to initiate the deployment process.

In this case, the operator has chosen to use a logistic regression. Refer to Figure 3-6.

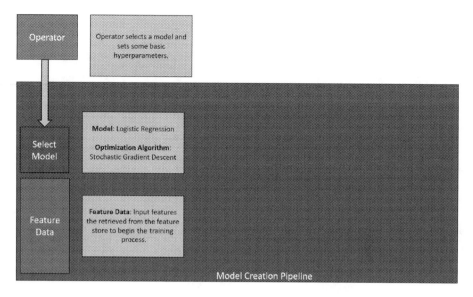

Figure 3-6. *A graphical representation of a pipeline where the operator has selected a logistic regression model. The rest of the steps have been hidden for now and will appear as we gradually move through the pipeline*

Data Preprocessing

With the model now selected and built, and with feature data supplied by the feature store, the process can now move forward to the next stage in the pipeline: data preprocessing. Refer to Figure 3-7.

115

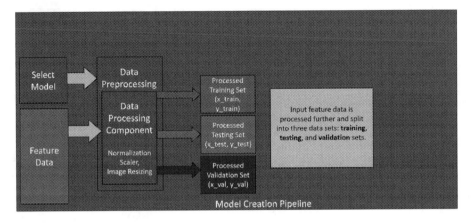

Figure 3-7. *The operator has chosen to normalize and resize the image data. The process creates a training set, a testing set, and a validation set*

The data preprocessing can be done manually or automatically. In this case, the data preprocessing only involves normalization and resizing of image feature data, so the operator can implement this manually. Depending on the level of automation, the operator can also call some function that takes in data and automatically processes it depending on the type of data and any other parameters provided.

Either way, the end of the processing stage will result in the data being broken up into subsets. In this example, the operator chose to create a training set, a testing set, and a validation set. Now, the operator can begin the training process.

Training Process

Depending on the framework being used, the operator can further split up the training data into a training set and a data validation set and use both in the training process. The data validation set exists totally separate from the training set (although it is derived from it) since the model never sees it during training. Its purpose is to periodically evaluate the model's performance on a data set that it has never seen before. Refer to Figure 3-8.

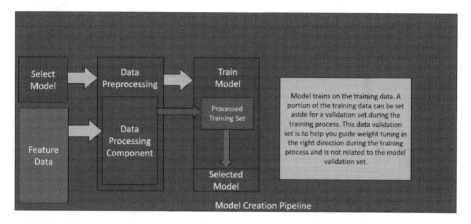

Figure 3-8. *The model training process begins*

In the context of deep learning, for example, the model can evaluate on the validation set at the end of each epoch, generating some metric data for the operator to see. Based on this, the operator can judge how the model is doing and whether or not it could be overfitting and adjust hyperparameters or model structure if needed.

The API can also be told what script to run in order to initiate this entire pipeline process. The script can contain the training, evaluation, and validation code all at once so the API can run this entire pipeline when needed.

Once the training process is done, the process moves to the evaluation stage.

Model Evaluation

In the evaluation stage, the model's performance is measured on a test data set that it has never seen. This performance will indicate to the operator whether or not the model is overfitting, especially if it performed extremely well in training but has trouble replicating those results in this stage. That is part of why the training data can be split to include some validation data, as it can be an early indicator of overfitting. This can be crucial especially if the model takes a significant amount of time to run. You would rather know earlier, partway through training, if the model is overfitting, rather than after it ran all night and is evaluated the next morning. Refer to Figure 3-9.

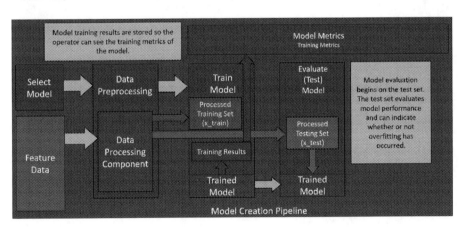

Figure 3-9. *Training results are stored in a common area (for example, the API could be called to monitor these results) for the metrics of the current model. Model evaluation begins on the trained model using the testing set*

Another thing to note again is that the validation stage could come before the evaluation stage, but in this case, the trained model will be evaluated first on a test data set before the validation stage begins. This is just to get a sense of how the model does on the testing set before hyperparameter tuning begins. Of course, hyperparameter tuning via the validation step could be performed first before the final model evaluation, but in some frameworks, model evaluation would come first. An example of this is a validation process like scikit-learn's cross-validation. Of course, you can evaluate the tuned model on the test set once again to get a final performance evaluation.

Once the evaluation finishes, metrics are stored by the API or by some other mechanism that the team has implemented, and the process moves on to the validation stage.

Model Validation

In this stage, the model begins the validation process, which attempts to seek the best hyperparameters. You could combine the use of a script to iterate through various configurations of hyperparameter values and utilize k-fold cross-validation, for example, to help decide the best hyperparameters. Refer to Figure 3-10.

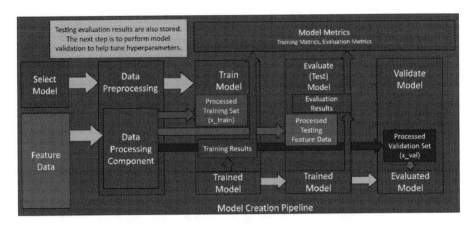

Figure 3-10. *Evaluation metrics are stored along with the training metrics by the API, and the validation process begins*

In any case, the point of a validation set is to help tune the model's hyperparameters. The team could even automate this process entirely if they tend to train a lot of models of the same few types, saving time and resources in the long run by automating the validation and hyperparameter tuning process for that set of models.

Finally, once the model achieves a good level of performance and finishes the validation stage, the validation results are stored, and all relevant data is displayed as a summary to the operator. Again, depending on the level of automation, perhaps the model is retrained and evaluated on the best hyperparameter setup discovered in the validation stage. The API simply needs to be told what metrics to track and it will automatically do so.

Model Summary

At this point, the operator can compare the outcome of this experiment
with that of other models, using the metrics as baselines for comparison.
The API can track the relevant metrics for different model runs and can
compare them all at the same time. Should the operator decide to move
forward with this particular model, the API and the MLOps software
can allow for deployment on a simple click of a button. Usually, the
deployment is to a staging environment first, where the functionality can
be tested further before moving directly into the production environment.
Everything is configurable, and the API can adapt to the needs of the
business and its workflow. If the developers want to deploy straight to
production, sure, though that could potentially be unwise considering the
case of failure. Refer to Figure 3-11.

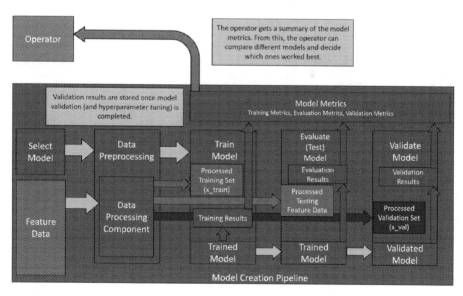

Figure 3-11. *Validation is complete, and all metrics are displayed to
the operator*

After the model passes the tests in the staging environment, it can then be deployed to the production environment, where it can be further monitored by the software.

Hopefully now you have a better understanding of what a pipeline really is. The pipelines for models and pipeline integration testing are similar, except they are assisted by MLOps software and APIs such as Databricks and MLFlow, for example. Let's now look at how you can go about using those APIs and software to help you implement MLOps.

How to Implement MLOps

MLOps sounds great. It helps you deploy machine learning models rapidly and helps maintain them once they're deployed. However, the biggest problem now seems to be the question of how to get there. The level of automation described in the setups requires significant work from both the "ML" and "Ops" sides of the workflow to achieve it. It almost seems better in the short run to build and deploy the models manually rather than devote resources to setting up the entire infrastructure, but this is simply unsustainable in the long run.

Also, Jupyter is great for performing experiments, so is there a way to track them as well? This sort of functionality would be extremely useful especially when teams are implementing advanced machine learning architectures from scratch, as it would let them compare the new models across all of the relevant metrics with deployed models or current architectures. Tasks like these are more convenient to do in a notebook and having to convert everything to a proper model file is simply further work.

The takeaway here is that accounting for these factors and more would require significant resources to plan, develop, and test. For smaller-scale businesses, this is an undertaking that's possibly beyond their reach. So, what now?

The good news is that are a great assortment of tools available to use now that essentially implement all of the automation for you, such as the API we looked at in the pipeline example earlier. Several examples of such tools that we will explore in later chapters are **MLFlow, Databricks**, **AWS SageMaker**, **Microsoft Azure**, **Google Cloud**, and **Datarobots**. With these tools, implementing MLOps principles into your workflow will be significantly easier.

In the case of MLFlow, integrating it into code is extremely simple. You only have to write a couple lines of code to track all of the metrics you need. The functionality of the API we looked at earlier in the pipeline example is all provided by MLFlow. Furthermore, MLFlow also saves the model for you, allowing for model serving functionality where given some data, the model returns its predictions.

MLFlow also integrates into Databricks, AWS SageMaker, Microsoft Azure, and can be deployed to Google Cloud as well, all of which are tools that help manage your MLOps setup and serve as platforms to deploy your models on. While the cloud platforms do provide some MLOps functionality, with the extent of this varying for each platform, the advantage of using MLFlow is that it lets you have the freedom of choice when it comes to one platform to commit to. Furthermore, it gives you a greater degree of freedom, as you can perform all the experiments locally and offline, and you can support models from many different frameworks. MLFlow also provides functionality to help you modularize any custom-built models or models made from other frameworks not explicitly supported.

And so, to really answer the question of how to implement MLOps, you will get familiar with MLFlow and explore each of those tools. The goal is to take the model we built in Chapter 2 all the way to deployment and beyond.

Summary

MLOps is a set of principles and practices adopted from DevOps and applied to machine learning. You explored three different types of MLOps setups with varying degrees of automation: **manual implementation**, **continuous model delivery**, and **continuous integration/continuous delivery of pipelines**. You identified that the manual implementation was riddled with issues regarding scalability and efficiency and you explored a setup that ensured continuous model delivery. Although this setup fixed many of the issues found in the manual setup, there were still some problems with pipeline integration testing to be solved. The final setup solved this issue too and ensured continuous integration and delivery of pipelines, completing the total automation setup.

You also looked into what a pipeline really is so that you can understand why they are so crucial to the automation setup. Finally, you learned about some tools that can help you implement MLOps into your workspace, avoiding the trouble of implementing all the automation from scratch. In the next chapter, you will look at MLFlow, an excellent API that lets you implement your own MLOps setups and is compatible with many platforms and frameworks.

CHAPTER 4

Introduction to MLFlow

In this chapter, we will cover what MLFlow is, what it does, and how you can implement MLOps setups into your existing projects. More specifically, we will cover how you can integrate MLFlow with scikit-learn, TensorFlow 2.0+/Keras, PyTorch, and PySpark. We will go over experiment creation; metric, parameter, and artifact logging; model logging; and how you can deploy models on a local server and query them for predictions.

Introduction

In the previous chapter, we went over what an optimal MLOps setup looks like. However, the level of automation presented would require an immense amount of resources dedicated to the project. Fortunately, there are APIs that do this for you, such as **MLFlow**. MLFlow is an API that allows you to integrate MLOps principles into your projects with minimal changes made to existing code. With just a couple lines of code here and there, you can track all of the details relevant to the project that you want. Furthermore, you can even save the model for future use in deployment, for example, and you can compare all of the metrics between individual models to help you select the best model.

© Sridhar Alla, Suman Kalyan Adari 2021
S. Alla and S. K. Adari, *Beginning MLOps with MLFlow*,
https://doi.org/10.1007/978-1-4842-6549-9_4

The great thing about MLFlow is that it abstracts everything for you. It packages and modularizes the models for you so that when you deploy the model and want to make predictions, all you need to do is simply pass in the input data in a certain format. All of the modularization that we discussed in the previous chapter with the pipelines is taken care of by MLFlow. MLFlow also allows you to create a wrapper around your model if your model prediction code needs to be different. We will look at this functionality in detail in the next chapter, when you deploy your models to Amazon SageMaker. Even with custom code, MLFlow will modularize it so that it will still work the same way as any other model once it is deployed and ready to make predictions.

In detail, we will go over the following in this chapter:

- **Creating experiments:** Experiments in MLFlow essentially allow you to group your models and any relevant metrics. For example, you can compare models that you've built in TensorFlow and in PyTorch and name this experiment something like `pytorch_tensorflow`. In the context of anomaly detection, you can create an experiment called `model_prototyping` and group all of the models that you want to test by running the training pipelines after setting `model_prototyping` as the experiment name.

  As you'll see shortly, grouping model training sessions by experiment can really help organize your workspace because you'll get a clear idea of the context behind trained models.

- **Model and metric logging:** MLFlow allows you to save a model in a modularized form and log all of the metrics related to the **model run**. A model run can be thought of as the model training, testing, and validation pipeline from the previous chapter. In MLFlow, you can mark the start and the end of each run and decide which metrics you want to save. Additionally, you can save graphs, so you can also view plots like confusion matrices and ROC curves. A **model run** is basically the instance in which MLFlow executes the code that you tell it to, so if you want, you can only train the model and leave it at that.

It is possible for you to train, evaluate, and even validate your model, logging all of the metrics for each respective step in the whole process. MLFlow gives you a lot of flexibility in how you define a model run. You can end the run after simply training it, or you can end the run after training and evaluating it. If you wish, you can even set up an entire validation script to log the entire process for you, allowing you to much more easily compare different hyperparameter setups all at once in MLFlow. We will explore how to perform model validation with MLFlow shortly when we revisit the scikit-learn experiment from Chapter 2.

- **Comparing model metrics:** MLFlow also allows you to compare different models and their metrics all at once. And so, when performing validation to help tune a model's hyperparameters, you can compare all of the selected metrics together in MLFlow using its user interface. In the previous chapter, you printed out everything, making the cell output possibly very large if the script is quite involved in its hyperparameter setups.

- **Model Registry:** MLFlow also adds functionality to allow you to implement a model registry, allowing you to define what stage a particular model is in. Databricks integrates quite well with MLFlow, providing built-in model registry functionality. You will explore how to use the MLFlow Model Registry when you look at Databricks in Appendix.

- **Local deployment:** MLFlow also allows you to deploy on a local server, allowing you to test **model inference**. Model inference is basically the prediction process of a model. Data is sent to the model in one of several standardized formats, and MLFlow returns the predictions made by the model.

  Such a setup can easily be converted to work on a hosted server as well. As you will see in the next several chapters, MLFlow also allows you to deploy your models on popular cloud services such as Amazon SageMaker, Microsoft Azure, Google Cloud, and Databricks. The process at its core remains similar to how you will perform local model serving. The only difference comes with where you host the model and the particular procedure for querying it.

With that being said, let's get started by revisiting the scikit-learn logistic regression model and integrating MLFlow into it.

MLFlow with Scikit-Learn

Before we begin, here are the versions of Python and the packages that were used:

- **Python**: 3.6.5

- **numpy**: 1.18.5

- **scikit-learn**: 0.22.2.post1

- **pandas**: 1.1.0

- **Matplotlib**: 3.2.1

- **Seaborn**: 0.10.1

- **MLFlow**: 1.10.0

You don't need the exact versions of the packages we used, but in case some functionality is removed, renamed, or just changed in the newer versions and the code runs into an error, you have the exact version of the module you can try running the code with.

MLFlow in particular is updated quite frequently, so you are more likely to run into issues running code with something like MLFlow compared to a package like numpy.

With that being said, let's dive into the first example. In this case, let's revisit the scikit-learn code from the previous chapter and add MLFlow integration to it.

Data Processing

First, you begin with all of the imports:

```
import numpy as np
import pandas as pd
import matplotlib #
```

```
import matplotlib.pyplot as plt
import seaborn as sns
import sklearn #
from sklearn.linear_model import LogisticRegression
from sklearn.model_selection import train_test_split
from sklearn.preprocessing import StandardScaler
from sklearn.metrics import roc_auc_score, plot_roc_curve,
confusion_matrix
from sklearn.model_selection import KFold

import mlflow
import mlflow.sklearn

print("Numpy: {}".format(np.__version__))
print("Pandas: {}".format(pd.__version__))
print("matplotlib: {}".format(matplotlib.__version__))
print("seaborn: {}".format(sns.__version__))
print("Scikit-Learn: {}".format(sklearn.__version__))
print("MLFlow: {}".format(mlflow.__version__))
```

The output should look something like Figure 4-1.

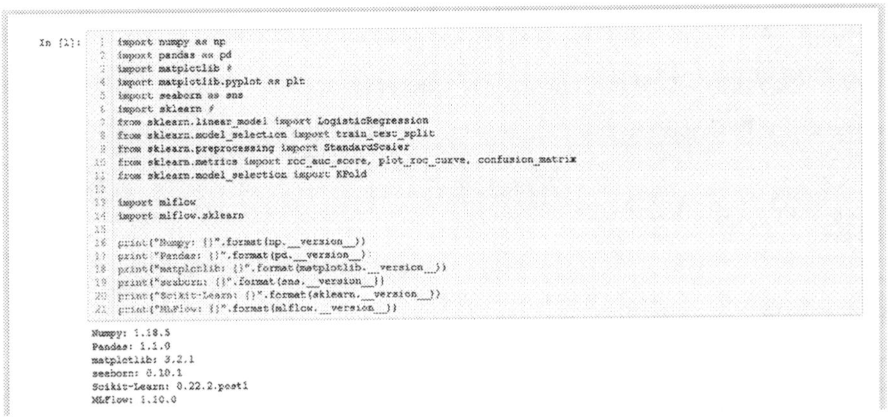

Figure 4-1. *The output of importing the necessary modules and printing out their versions*

Now you can move on to loading the data:

```
data_path = "data/creditcard.csv"

df = pd.read_csv(data_path)
df = df.drop("Time", axis=1)
```

Refer to Figure 4-2 to see the code in a cell.

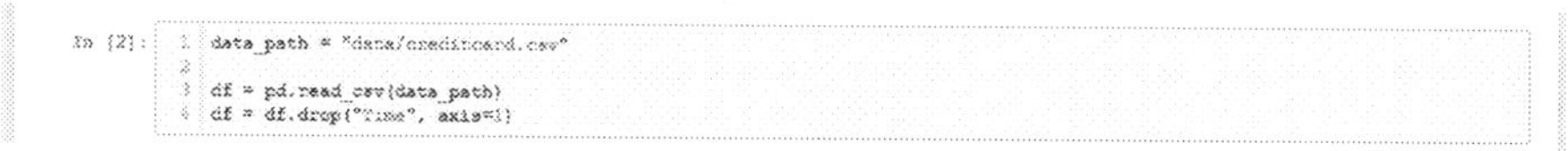
```
In [2]:
data_path = "data/creditcard.csv"

df = pd.read_csv(data_path)
df = df.drop("Time", axis=1)
```

Figure 4-2. *Loading the data set and dropping the column named Time because it adds very large data values that ultimately don't have much of a correlation with the column Class. Model performance is boosted slightly simply by dropping the extraneous information*

Note that you are once again dropping the column Time.

You can now check to see if the data loaded in correctly:

```
df.head()
```

Refer to Figure 4-3 to see the head()function.

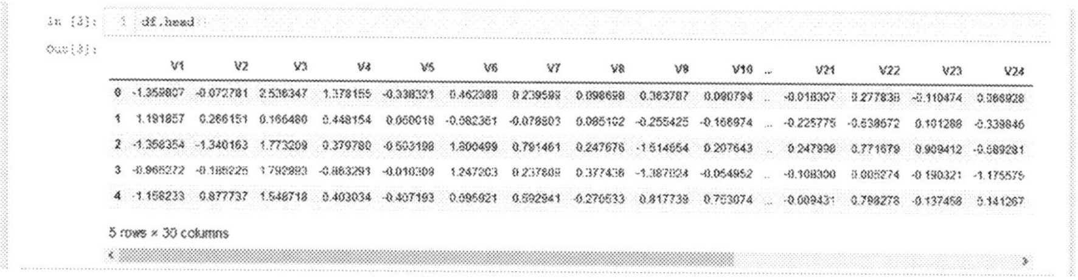
```
In [3]: df.head()
Out[3]:
```

	V1	V2	V3	V4	V5	V6	V7	V8	V9	V10	...	V21	V22	V23	V24
0	-1.359807	-0.072781	2.536347	1.378155	-0.338321	0.462388	0.239599	0.098698	0.363787	0.090794	...	-0.018307	0.277838	-0.110474	0.066928
1	1.191857	0.266151	0.166480	0.448154	0.060018	-0.082361	-0.078803	0.085102	-0.255425	-0.166974	...	-0.225775	-0.638672	0.101288	-0.339846
2	-1.358354	-1.340163	1.773209	0.379780	-0.503198	1.800499	0.791461	0.247676	-1.514654	0.207643	...	0.247998	0.771679	0.909412	-0.689281
3	-0.966272	-0.185226	1.792993	-0.863291	-0.010309	1.247203	0.237609	0.377436	-1.387024	-0.054952	...	-0.108300	0.005274	-0.190321	-1.175575
4	-1.158233	0.877737	1.548718	0.403034	-0.407193	0.095921	0.592941	-0.270533	0.817739	0.753074	...	-0.009431	0.798278	-0.137458	0.141267

5 rows × 30 columns

Figure 4-3. *Verifying that the data was loaded correctly by using the head() function. As you can see, the columns and the data have loaded in correctly*

Again, you are dropping the column Time from the data frame this time. This is because this column was found to add data that isn't very helpful in finding an anomaly and only adds extra complexity to the data.

In the case of deep learning models, your model might eventually learn that the Time data does not correlate very well with the Class labels and may place less importance on nodes processing that data. Eventually, it might even ignore the Time data. However, you can speed up the learning process by cutting out these types of features from your training sets. This is because you're sparing the models the time and resources needed to figure that out.

Moving on, you will split the normal points and the anomalies:

```
normal = df[df.Class == 0].sample(frac=0.5,
random_state=2020).reset_index(drop=True)
anomaly = df[df.Class == 1]
```

Let's print out their respective shapes:

```
print(f"Normal: {normal.shape}")
print(f"Anomaly: {anomaly.shape}")
```

Refer to Figure 4-4 to see the above two cells in Jupyter along with their outputs.

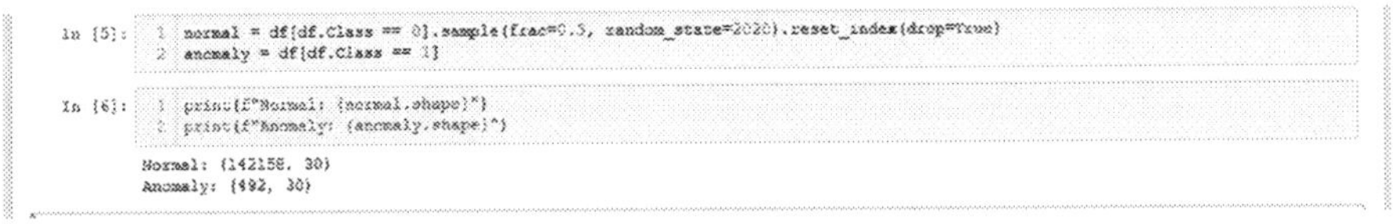

Figure 4-4. *Randomly sampling 50% of all the normal data points in the data frame and picking out all of the anomalies from the data frame as separate data frames. Then, you print the shapes of both data sets. As you can see, the normal points massively outnumber the anomaly points*

You are going to split the normal and anomaly sets into train-test-validate subsets. Run the following two code blocks:

```
normal_train, normal_test = train_test_split(normal,
test_size = 0.2, random_state = 2020)
anomaly_train, anomaly_test = train_test_split
(anomaly, test_size = 0.2, random_state = 2020)

normal_train, normal_validate = train_test_split(normal_train,
test_size = 0.25, random_state = 2020)
anomaly_train, anomaly_validate = train_test_split
(anomaly_train, test_size = 0.25, random_state = 2020)
```

Refer to Figure 4-5 to see both code blocks in their respective cells.

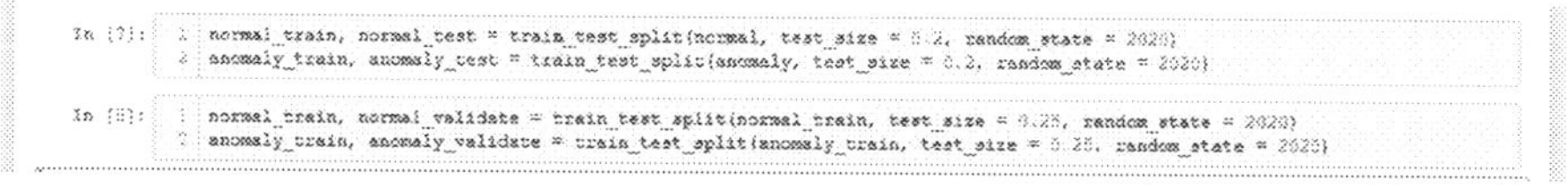

Figure 4-5. *Partitioning the normal and anomaly data frames separately into train, test, and validation splits. Initially, 20% of the normal and anomaly points are used as the test split. From the remaining 80% of data, 25% of that train split is used as the validation split, meaning the validation split is 20% of the original data. This leaves the final training split at 60% of the original data. In the end, the train-test-validate split has a 60-20-20 ratio, respectively*

Now, you can process these sets and create the x-y splits:

```
x_train = pd.concat((normal_train, anomaly_train))
x_test = pd.concat((normal_test, anomaly_test))
x_validate = pd.concat((normal_validate, anomaly_validate))

y_train = np.array(x_train["Class"])
y_test = np.array(x_test["Class"])
y_validate = np.array(x_validate["Class"])
```

```
x_train = x_train.drop("Class", axis=1)
x_test = x_test.drop("Class", axis=1)
x_validate = x_validate.drop("Class", axis=1)
```

Refer to Figure 4-6 to see the above code block in a cell.

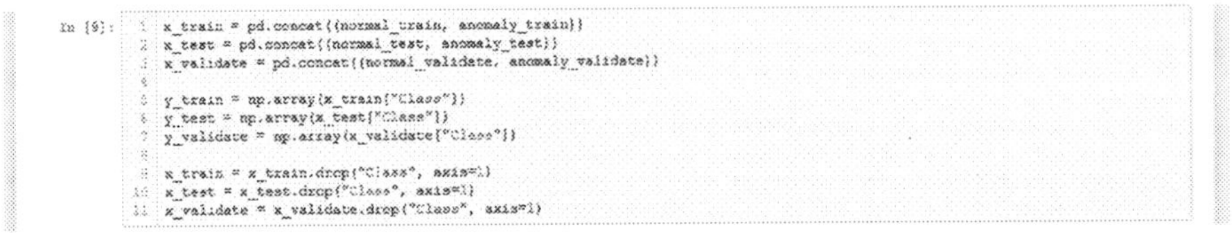

Figure 4-6. *Creating the respective x and y splits of the training, testing, and validation sets by concatenating the respective normal and anomaly sets. You drop Class from the x-sets because it would be cheating otherwise to give it the label directly. You are trying to get the model to learn the labels by reading the x-data, not learn how to read the Class column in the x-data*

You can print out the shapes of these sets:

```
print("Training sets:\nx_train: {} \ny_train:
{}".format(x_train.shape, y_train.shape))
print("\nTesting sets:\nx_test: {} \ny_test:
{}".format(x_test.shape, y_test.shape))
print("\nValidation sets:\nx_validate: {} \ny_validate: {}".
format(x_validate.shape, y_validate.shape))
```

Refer to Figure 4-7 to see the output shapes.

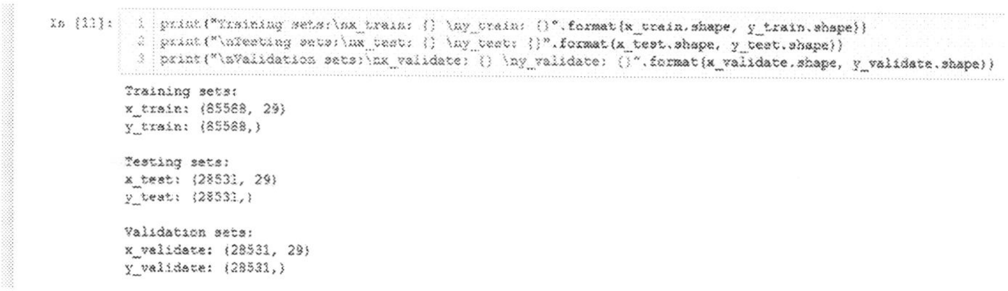

```
In [11]:  1  print("Training sets:\nx_train: {} \ny_train: {}".format(x_train.shape, y_train.shape))
          2  print("\nTesting sets:\nx_test: {} \ny_test: {}".format(x_test.shape, y_test.shape))
          3  print("\nValidation sets:\nx_validate: {} \ny_validate: {}".format(x_validate.shape, y_validate.shape))

          Training sets:
          x_train: (85588, 29)
          y_train: (85588,)

          Testing sets:
          x_test: (28531, 29)
          y_test: (28531,)

          Validation sets:
          x_validate: (28531, 29)
          y_validate: (28531,)
```

Figure 4-7. *Printing out the shapes of the training, testing, and validation sets*

Finally, you scale your data using scikit-learn's standard scaler:

```
scaler = StandardScaler()
scaler.fit(pd.concat((normal, anomaly)).drop("Class", axis=1))

x_train = scaler.transform(x_train)
x_test = scaler.transform(x_test)
x_validate = scaler.transform(x_validate)
```

Refer to Figure 4-8.

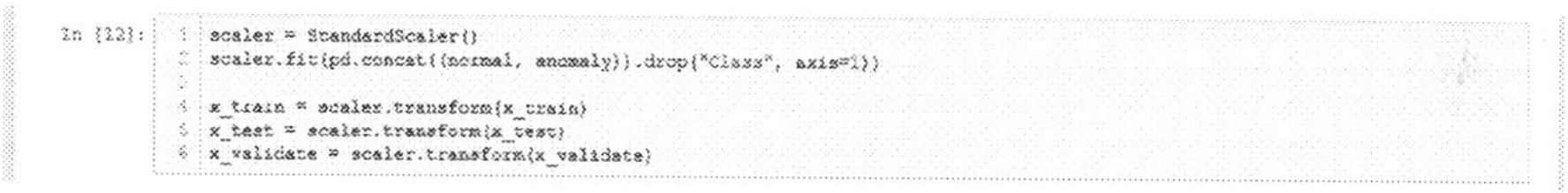

```
In [12]:  1  scaler = StandardScaler()
          2  scaler.fit(pd.concat((normal, anomaly)).drop("Class", axis=1))

          4  x_train = scaler.transform(x_train)
          5  x_test = scaler.transform(x_test)
          6  x_validate = scaler.transform(x_validate)
```

Figure 4-8. *Fitting the scaler on the superset of normal and anomaly points after dropping Class to scale the x-sets*

Training and Evaluating with MLFlow

All that is left now is to train and evaluate your model. We will showcase validation with MLFlow functionality in a bit, but first let's define the train and test functions to organize the code. This is also where you start integrating MLFlow into your code. Here is the train function:

```
def train(sk_model, x_train, y_train):
    sk_model = sk_model.fit(x_train, y_train)

    train_acc = sk_model.score(x_train, y_train)
    mlflow.log_metric("train_acc", train_acc)

    print(f"Train Accuracy: {train_acc:.3%}")
```

Refer to Figure 4-9 to see this code in a cell.

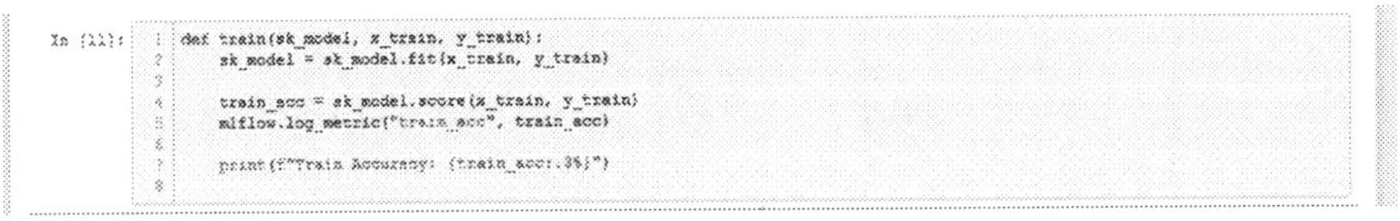

Figure 4-9. *Defining the train function to better organize the code. Additionally, you are defining a training accuracy metric that will be logged by MLFlow*

You may have noticed the first of the new code with this line:

```
mlflow.log_metric("train_acc", train_acc)
```

You create a new metric here specifically for the training accuracy so that you can keep track of this metric. Furthermore, you are telling MLFlow to log this metric, so that MLFlow will keep track of this value in each run. When you log multiple runs, you can compare this metric across each of those runs so that you can pick a model with the best AUC score for example.

Let's now move on to the evaluate function:

```python
def evaluate(sk_model, x_test, y_test):
    eval_acc = sk_model.score(x_test, y_test)

    preds = sk_model.predict(x_test)
    auc_score = roc_auc_score(y_test, preds)

    mlflow.log_metric("eval_acc", eval_acc)
    mlflow.log_metric("auc_score", auc_score)

    print(f"Auc Score: {auc_score:.3%}")
    print(f"Eval Accuracy: {eval_acc:.3%}")

    roc_plot = plot_roc_curve(sk_model, x_test, y_test,
    name='Scikit-learn ROC Curve')
    plt.savefig("sklearn_roc_plot.png")
    plt.show()
    plt.clf()

    conf_matrix = confusion_matrix(y_test, preds)
    ax = sns.heatmap(conf_matrix, annot=True,fmt='g')
    ax.invert_xaxis()
    ax.invert_yaxis()
    plt.ylabel('Actual')
    plt.xlabel('Predicted')
    plt.title("Confusion Matrix")
    plt.savefig("sklearn_conf_matrix.png")

    mlflow.log_artifact("sklearn_roc_plot.png")
    mlflow.log_artifact("sklearn_conf_matrix.png")
```

Refer to Figure 4-10 to see the above code in a cell.

```
In [12]:    1  def evaluate(sk_model, x_test, y_test):
            2
            3      eval_acc = sk_model.score(x_test, y_test)
            4
            5      preds = sk_model.predict(x_test)
            6      auc_score = roc_auc_score(preds, y_test)
            7
            8      mlflow.log_metric("eval_acc", eval_acc)
            9      mlflow.log_metric("auc_score", auc_score)
           10
           11      print(f"Auc Score: {auc_score:.3%}")
           12      print(f"Eval Accuracy: {eval_acc:.3%}")
           13
           14      roc_plot = plot_roc_curve(sk_model, x_test, y_test, name='Scikit-learn ROC Curve')
           15      plt.savefig("sklearn_roc_plot.png")
           16      plt.show()
           17      plt.clf()
           18
           19      conf_matrix = confusion_matrix(y_test, preds)
           20      ax = sns.heatmap(conf_matrix, annot=True, fmt='g')
           21      ax.invert_xaxis()
           22      ax.invert_yaxis()
           23      plt.ylabel('Actual')
           24      plt.xlabel('Predicted')
           25      plt.title("Confusion Matrix")
           26      plt.savefig("sklearn_conf_matrix.png")
           27
           28      mlflow.log_artifact("sklearn_roc_plot.png")
           29      mlflow.log_artifact("sklearn_conf_matrix.png")
```

Figure 4-10. *A function to calculate the evaluation metrics for the AUC score and accuracy. Plots for the confusion matrix and the ROC curve are generated, and both the metrics and the graphs are logged to MLFlow*

Once again, you have told MLFlow to log two more metrics: the AUC score and the accuracy on the test set. You do so with these lines of code:

```
mlflow.log_metric("eval_acc", eval_acc)
mlflow.log_metric("auc_score", auc_score)
```

Furthermore, you can also tell MLFlow to save the plots generated by matplotlib and by seaborn. With this, you can look at each of the graphs for each training run and do so in a highly organized manner. You must first save these plots, which you do in the same directory. Then, you must tell MLFlow to grab the artifacts to log them like so:

```
mlflow.log_artifact("sklearn_roc_plot.png")
mlflow.log_artifact("sklearn_conf_matrix.png")
```

Make sure that they have the same names as the graphs you saved.

Logging and Viewing MLFlow Runs

Finally, let's run the code that actually sets the experiment name, starts the MLFlow run, and executes all this code:

```
sk_model = LogisticRegression(random_state=None,
max_iter=400, solver='newton-cg')

mlflow.set_experiment("scikit_learn_experiment")
with mlflow.start_run():
    train(sk_model, x_train, y_train)
    evaluate(sk_model, x_test, y_test)
    mlflow.sklearn.log_model(sk_model, "log_reg_model")
    print("Model run: ", mlflow.active_run().info.run_uuid)
mlflow.end_run()
```

Notice the new lines of MLFlow code. We will go through them one by one.

First, let's begin with what appears to set the experiment name:

```
mlflow.set_experiment("scikit_learn_experiment")
```

What this does is that it puts the run under whatever experiment name you pass in as a parameter. If that name does not exist, MLFlow will create a new one under that name and put the run there.

```
with mlflow.start_run():
    ...
    ...
```

This line of code allows you to chunk all of your code under the context of one MLFlow run. This ensures that there are no discrepancies between where your metrics are being logged, and that it doesn't create two different runs when you mean it to log everything for the same run.

```
mlflow.sklearn.log_model(sk_model, "log_reg_model")
```

139

This line of code is the general convention to use when you're logging a model. The parameters, in order, are the model you're saving and then the name you're setting for the model when saving. In this case, you are saving your logistic regression model with the name `log_reg_model` in this run.

As you will see later, most other frameworks follow the same style when saving the model. There are a couple exceptions, but we will cover this when the time comes. In this case, you are calling `mlflow.sklearn`, but if you wanted to log a PySpark model, you would do `mlflow.spark`.

Basically, the framework the model was built in must match the framework module of MLFlow when logging the model. It is possible to create a custom "model" in MLFlow and log this as well, something that is covered in the documentation. You can use this custom model to then specify how you want the prediction function to work. If you'd like to process the data some more before making predictions, for example, MLFlow allows you to specify this extra functionality through the use of the MLFlow PyFunc module. Refer to the documentation, which you can find here: `www.mlflow.org/docs/latest/models.html#model-customization`.

```
print("Model run: ", mlflow.active_run().info.run_uuid)
```

This line of code essentially gets the current run that the model and metrics are being logged to and prints it out. This makes it handy if you want to retrieve the run directly from the notebook itself instead of going to the UI to do so.

```
mlflow.end_run()
```

Finally, this tells MLFlow to end the current run. In cases where there is an error in the MLFlow start run code block, and the run does not terminate, do this to forcibly end the current run. Basically, it is there to ensure that MLFlow stops the run after you executed all the code relevant to the current run.

Moving on, refer to Figure 4-11 to see the full output of the code.

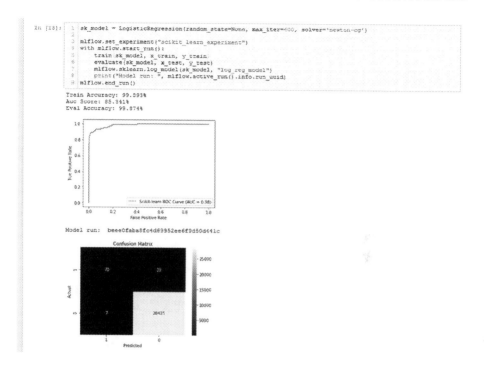

Figure 4-11. *The output of running the MLFlow experiment. Under an MLFlow run context, you are training the model, outputting the graphs from the evaluation function, and logging all the metrics including the model to this run*

You can see that MLFlow automatically generates a new experiment if it does not already exist, so you can create a new experiment directly from the code. You can also see that the rest of the code basically outputs as usual, except it also prints the run ID of the current MLFlow run just as you specified. You will use this later when you select the specific model that you want to serve. What you will do next is open up the UI MLFlow provides where you can actually look at all the experiments and model runs. Finally, you also log the model itself as an artifact with MLFlow. MLFlow will modularize this code so that it will work with the code provided by MLFlow to support implementations of a variety of MLOps principles.

The following was done on **Windows 10**, but it should be the same on MacOS or Linux. First, open command prompt/powershell/terminal. Then, you must go into the directory that contains this notebook file. List the contents of the directory (or view this in file explorer/within Jupyter itself) and you will notice a new directory named mlruns.

If you installed all of your packages in Conda, make sure you've activated the Conda environment before running this.

What you want to do now is to make sure your command prompt, powershell, or terminal is in the same directory that contains mlruns, and type the following:

```
mlflow ui -p 1234
```

The command mlflow ui hosts the MLFlow UI locally on the default port of 5000. However, the options -p 1234 tell it that you want to host it specifically on the port 1234.

If it all goes well, and it can take several seconds, you should see something like Figure 4-12.

```
(p36) C:\Users\Shumpu\work\Books\2020 MLOps\Chapter 4>ls
1                              'MLFlow PySpark.ipynb'  'MLFlow TensorFlow.ipynb'   sklearn_conf_matrix.png
'MLFlow Keras.ipynb'           'MLFlow PyTorch.ipynb'   data                        sklearn_roc_plot.png
'MLFlow Local Serving.ipynb'  'MLFlow Sklearn.ipynb'   mlruns

(p36) C:\Users\Shumpu\work\Books\2020 MLOps\Chapter 4>mlflow ui -p 1234
c:\users\shumpu\anaconda2\envs\p36\lib\site-packages\waitress\adjustments.py:445: DeprecationWarning: In future versions
 of Waitress clear_untrusted_proxy_headers will be set to True by default. You may opt-out by setting this value to Fals
e, or opt-in explicitly by setting this to True.
  DeprecationWarning,
Serving on http://kubernetes.docker.internal:1234
```

Figure 4-12. *Making sure that the current directory contains the folder mlruns and calling the command to start the UI. If successful, it should state "Serving on http:// … :1234." We have docker on our system, hence why yours might say localhost instead of kubernetes. docker.internal*

Now, open a browser and type in `http://localhost:1234` or `http://127.0.0.1:1234`. Both should take you to the same MLFlow UI. If you used a different port, it should generally look like this:

`http://localhost:PORT_NUMBER` or `http://127.0.0.1:PORT_NUMBER`, where you replace `PORT_NUMBER` with the one you used. If you did not specify a port parameter, then the default port used by MLFlow is 5000.

Regardless, if it works correctly, you should see something like Figure 4-13 once you visit that URL.

Figure 4-13. *Your MLFlow UI should look something like this. To the left are the experiments. Notice that there is an experiment titled Default and one titled scitkit_learn_experiment, which is the one you just created*

Notice that there is now an experiment titled `scikit_learn_experiment`. Click it, and you should see something like Figure 4-14.

You can see the run that just completed, along with the metrics you logged. Click it so that you can explore it. The run that was just completed should have a green check mark beside the time stamp when it finished if everything went well, which you can see is the case in Figure 4-14.

Figure 4-14. *This is what your experiment, scikit_learn_experiment, should look like once you click it. Notice that there is one run here, which is what was just created*

You should see something like Figure 4-15.

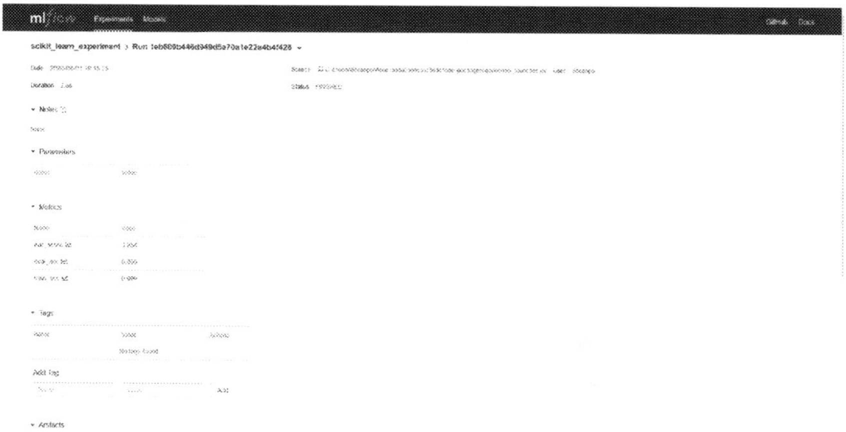

Figure 4-15. *This is the run that was just completed. Notice that the metrics you logged show up here*

You should now see the details of this run much more clearly. Here, you can see all of the parameters and metrics that were logged. Keep scrolling down and you should be able to see all of the logged artifacts. Refer to Figure 4-16.

Figure 4-16. *The logged artifacts of this run. Notice that the graphs appear to be logged as well as the model itself, which was named log_reg_model when you were logging it in the code*

Here, you can see the model that has been logged, along with the two graphs that you logged as artifacts. Click the graphs and you should see something like Figure 4-17.

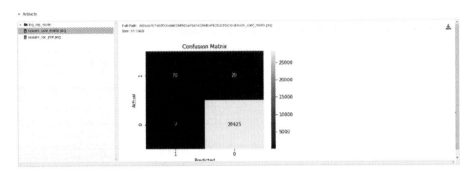

Figure 4-17. *Inspecting the graph of the confusion matrix that you saved. Feel free to click the other graph as well, which is of the ROC plot*

Amazing, right? Everything is extremely organized, and you don't have to worry about creating multiple folders for everything and staying organized. Simply tell MLFlow what to do and it will log all the information relevant to this run that you need. You can log your deep learning model's hyperparameters for learning rate, number of epochs, specific optimizer parameters like beta1 and beta2 for the Adam optimizer, and so on.

You can even log graphs, as you can see in Figure 4-17, along with the models themselves. With MLFlow, you can stay highly organized with your experiments even if you don't necessarily need the deployment capabilities to the cloud services.

Let's now try logging a few more runs. Rerun the cell in Figure 4-11 a couple times to completion and go back to the MLFlow UI. Make sure you have selected the experiment named scikit_learn_experiment. You should see something like Figure 4-18.

Figure 4-18. Revisiting your experiment after logging some runs in. The runs are logged in ascending order by timestamp, so the latest runs are on top

Let's compare the metrics you've logged for these runs. Select at least two runs, and ensure your UI looks somewhat like Figure 4-19. We selected three runs.

Figure 4-19. This is what your UI should look like after selecting several runs. Make sure to select at least two so that there is something to compare. Also notice that the button named Compare turns solid

147

After clicking Compare, you should see something like Figure 4-20.

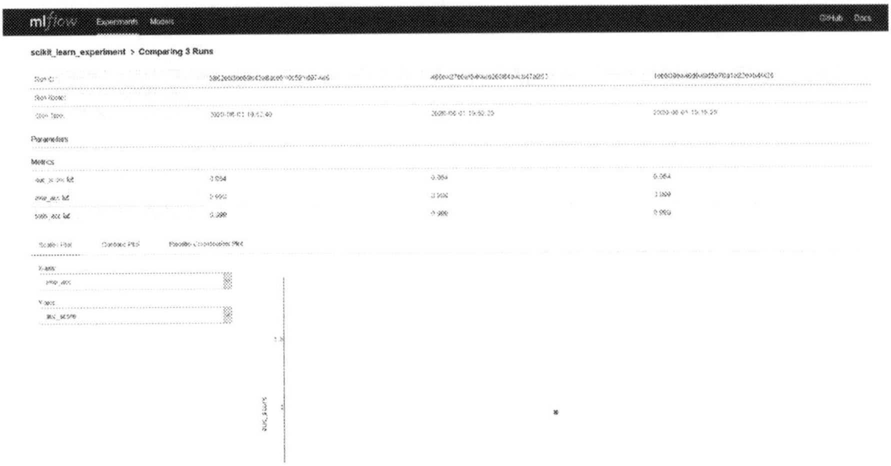

Figure 4-20. *The UI after selecting three runs to compare. As you can see, you can look at all of the metrics at once. There is also a graphing tool that lets you compare these values graphically, though you won't see proper graphs as every value is the same across the runs*

Here, you can directly compare the relevant parameters and metrics between the runs you have chosen. You have the option of viewing a scatter plot, a contour plot, or a parallel coordinates plot. Feel free to play around with the metrics and with the plots. You can even save these plots if you wish.

Note that since these runs have the exact same metrics, there will only appear to be one point plotted.

Loading a Logged Model

Next, let's briefly look at how you can load the models logged by MLFlow. Go back to the experiment and click a run. Note the run ID at the top and copy it. Then, go back to the notebook, and run the following. Note that there is a placeholder for the run ID:

```
loaded_model = mlflow.sklearn.load_model
("runs:/YOUR_RUNID_HERE/log_reg_model")
```

To better understand what this path is, let's split it up into three sections: the format (runs:/), the run ID (YOUR_RUNID_HERE), and the model name that you used when you logged it (log_reg_model).

In our case, our run ID was 3862eb3bd89b43e8ace610c521d974e6, so our cell looks like Figure 4-21. Ensure your code looks somewhat like Figure 4-21, with the only difference being the run ID that you chose since it will be different from ours.

Figure 4-21.   *The code to load a model that we logged using the specific run ID we logged it in along with the model's name we used when we logged it*

This is now the same model that you had when MLFlow logged it in the first place. With this, you can call something like .score() and see that it's the same as during training:

```
loaded_model.score(x_test, y_test)
```

This outputs the accuracy as the model is evaluated on the test set. If this truly is the same model, then the accuracy should match what was output earlier during the evaluation portion of the model run.

Refer to Figure 4-22 to see the output.

Figure 4-22.   *This is the evaluation accuracy of the loaded model after evaluation on the test sets. If you compare this with Figure 4-11, you can see that the numbers more or less match, disregarding rounding*

As you can see, this value matches the evaluation accuracy from Figure 4-11.

Now you know how to load a model from a specific MLFlow run.

With that, you've seen some of the functionality that MLFlow provides and how it can help in keeping your prototyping experiments much more organized. As you will see shortly, this entire pipeline that you just explored is pretty much all you need to recreate the train, test, validate pipeline that you saw earlier. Before you move on to looking at how you can use MLFlow with other frameworks, let's go over how you can use MLFlow functionality to vastly improve the model validation process.

Model Validation (Parameter Tuning) with MLFlow

Parameter Tuning – Broad Search

Just like in Chapter 2, you will use a script to help with model validation with respect to hyperparameter tuning. The tuning script will largely remain the same, except for a few modifications where MLFlow code has been added in.

Run the following code to set the range of anomaly weights and to set the number of folds:

```
anomaly_weights = [1, 5, 10, 15]
num_folds = 5
kfold = KFold(n_splits=num_folds, shuffle=True,
random_state=2020)
```

The code should look like Figure 4-23.

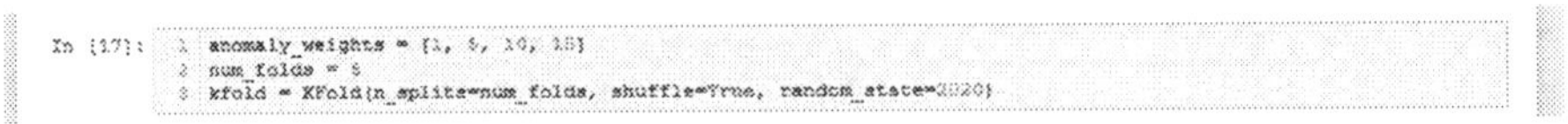

Figure 4-23. *The code to determine the list of anomaly weights to perform validation over, to determine the number of folds, and to initialize the KFolds generator based on the number of folds*

Now, paste the following. This is the first half of the entire function:

```python
mlflow.set_experiment("sklearn_creditcard_broad_search")
logs = []
for f in range(len(anomaly_weights)):
    fold = 1
    accuracies = []
    auc_scores= []
    for train, test in kfold.split(x_validate, y_validate):
        with mlflow.start_run():
            weight = anomaly_weights[f]
            mlflow.log_param("anomaly_weight", weight)

            class_weights= {
                0: 1,
                1: weight
            }
            sk_model = LogisticRegression(random_state=None,
                                max_iter=400,
                                solver='newton-cg',
                                class_weight=class_
                                weights).fit
                                (x_validate[train],
                                y_validate[train])

            for h in range(40): print('-', end="")
            print(f"\nfold {fold}\nAnomaly Weight: {weight}")

            train_acc = sk_model.score(x_validate[train],
            y_validate[train])
            mlflow.log_metric("train_acc", train_acc)

            eval_acc = sk_model.score(x_validate[test],
            y_validate[test])
            preds = sk_model.predict(x_validate[test])

            mlflow.log_metric("eval_acc", eval_acc)
```

Here is some more of the code. Make sure this all aligns with the code from above.

```python
        try:
            auc_score = roc_auc_score(y_validate[test], preds)
        except:
            auc_score = -1

        mlflow.log_metric("auc_score", auc_score)

        print("AUC: {}\neval_acc: {}".format(auc_score,
        eval_acc))

        accuracies.append(eval_acc)
        auc_scores.append(auc_score)

        log = [sk_model, x_validate[test],
        y_validate[test], preds]
        logs.append(log)
        mlflow.sklearn.log_model(sk_model,
        f"anom_weight_{weight}_fold_{fold}")

        fold = fold + 1
        mlflow.end_run()
print("\nAverages: ")
print("Accuracy: ", np.mean(accuracies))
print("AUC: ", np.mean(auc_scores))

print("Best: ")
print("Accuracy: ", np.max(accuracies))
print("AUC: ", np.max(auc_scores))
```

First, let's look at what that giant chunk of code looks like in a cell. Ensure your code and alignment matches Figure 4-24.

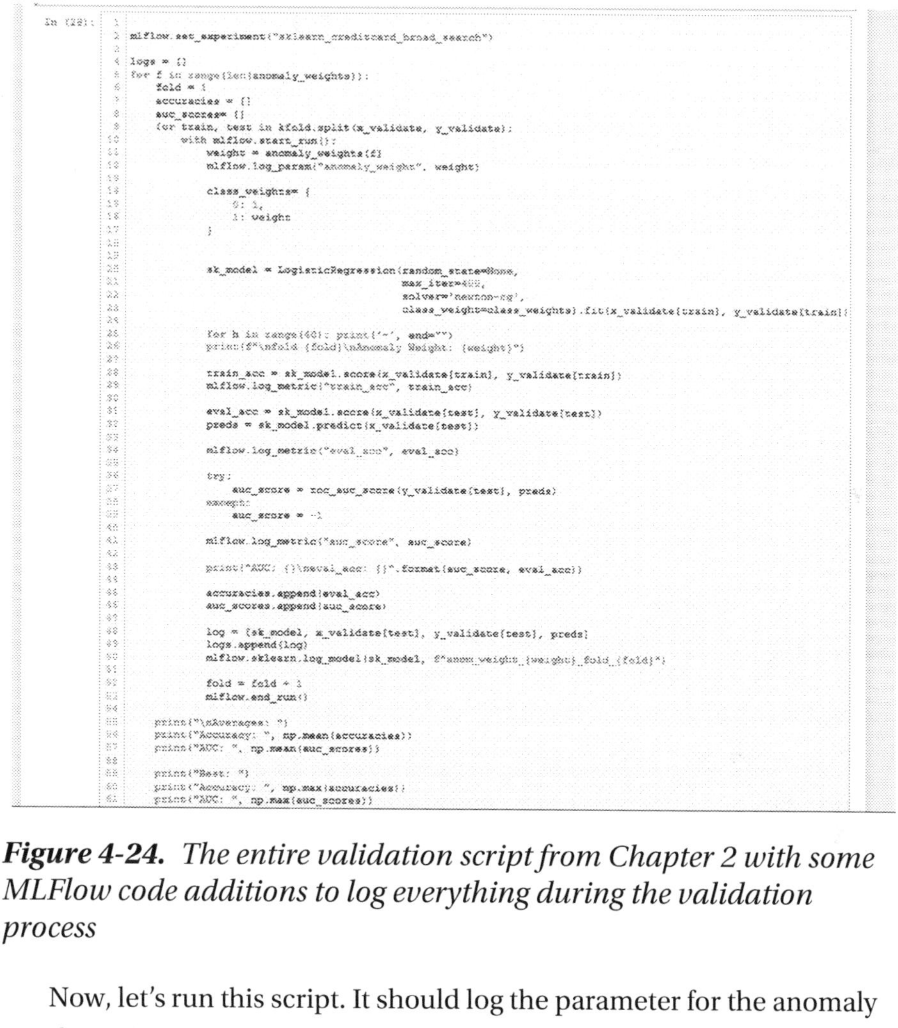

Figure 4-24. *The entire validation script from Chapter 2 with some MLFlow code additions to log everything during the validation process*

Now, let's run this script. It should log the parameter for the anomaly weight and all of the metrics that you specified for every fold generated. When the script finishes, go to your MLFlow UI and switch the experiment to `sklearn_creditcard_broad_search` to see all the runs you just logged. You should see something like in Figure 4-25.

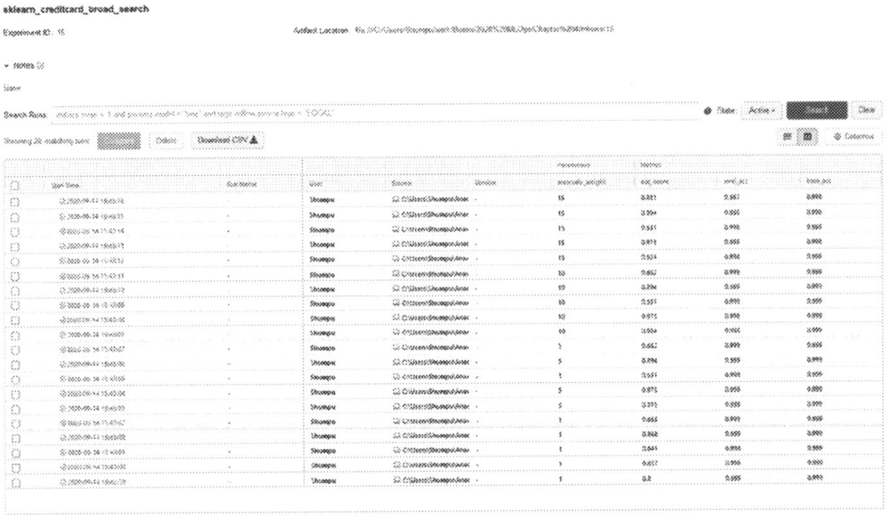

Figure 4-25. *The output you should see after the validation experiment has finished. Make sure you select the experiment titled sklearn_creditcard_broad_search*

Let's try sorting this by the AUC score to find the best parameters for the AUC. In the `metrics` column, click `auc_score`.

The action should result in something that looks like Figure 4-26.

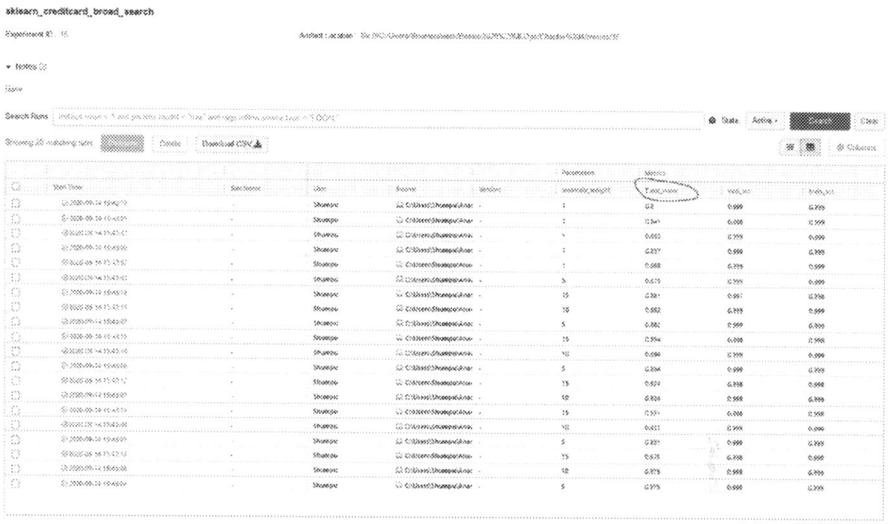

Figure 4-26. *The values are all sorted by auc_score in descending order. We've highlighted this column so that you can more easily spot the difference between this figure and Figure 4-25. As you can see, the AUC scores are in ascending order. You want to see the best AUC scores, so you must sort in descending order*

You want to sort the columns in descending order, so click it again to see something that looks like Figure 4-27.

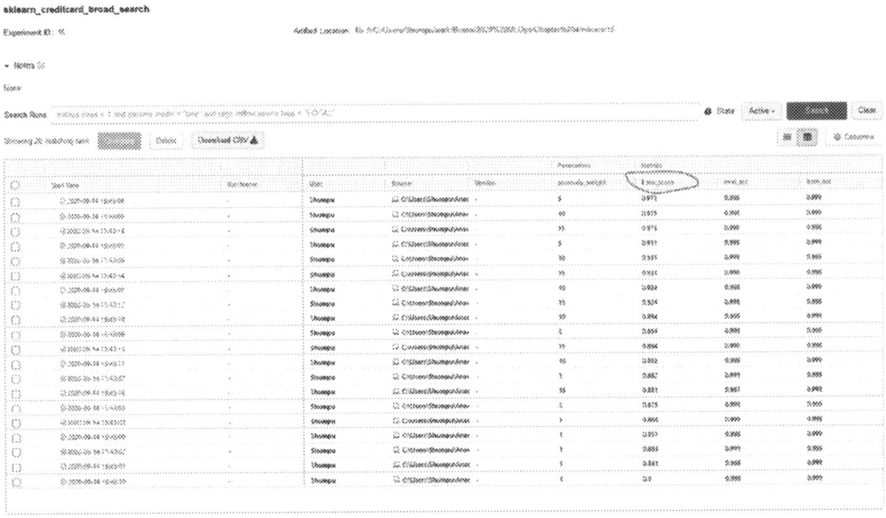

Figure 4-27. *The values are now sorted by AUC score in descending order. Now you can see the runs that produced the best AUC scores along with the specific anomaly weight it had in that run*

Perhaps you don't really care about anything but the absolute best scores. Say that you are targeting AUC scores that are at least 0.90. How would you go about filtering everything? Well, the UI provides a search bar that performs a search based on the SQL WHERE clause. So, to filter your output, type the following and click Search:

```
metrics."auc_score" >= 0.90
```

You should see something like Figure 4-28. If you have copied and pasted the line of code, be sure to delete it and put in the quotation marks again if you encounter any errors about the quotation marks.

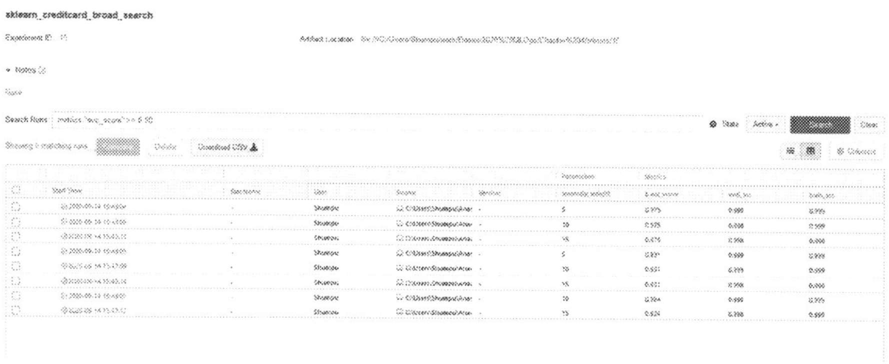

Figure 4-28. *The results of filtering all of the AUC scores to be above 0.90. As you can see, only a handful of runs produced AUC scores that are at least 0.90*

Notice that we put "auc_score" in quotation marks. This is for cases where the metric name that you've logged contains characters like a dash where it might not recognize the name if you typed it out like so:

```
metrics.auc-score
```

The proper convention for a metric logged as "auc-score" would be to filter it like so:

```
metrics."auc-score" >= 0.90
```

Now let's say that of these scores, you want to see the scores for anomaly weights of 5 only. It doesn't appear that there are any results with the anomaly weight of 1, so we will start with 5. For that, let's type and search the following:

```
metrics."auc_score" >= 0.90 AND parameters.anomaly_weight = "5"
```

You should see something like Figure 4-29.

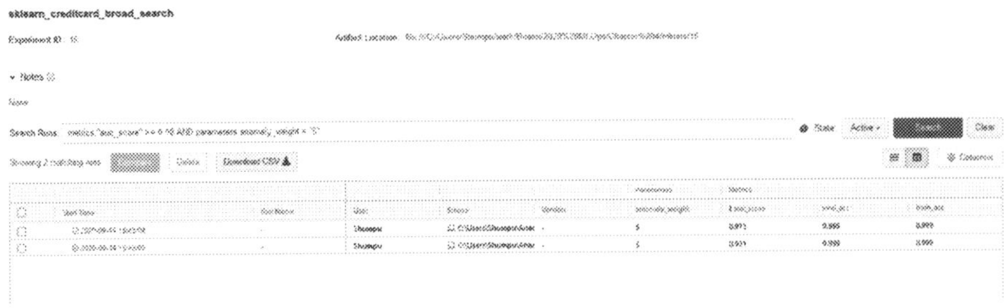

Figure 4-29. *Filtering the runs to have only runs with the anomaly weight set to 5 and to have an AUC score above 0.90*

You put the 5 in quotation marks because the parameters seem to be logged as string values, whereas the metrics are logged as floats.

From this output, it seems that only two of the five folds with the anomaly weight set to 5 had an AUC score above 0.90. Let's quickly search over the other parameters and check how many folds had an AUC score above 0.90 as well.

For filtering the anomaly weight by 10, refer to Figure 4-30.

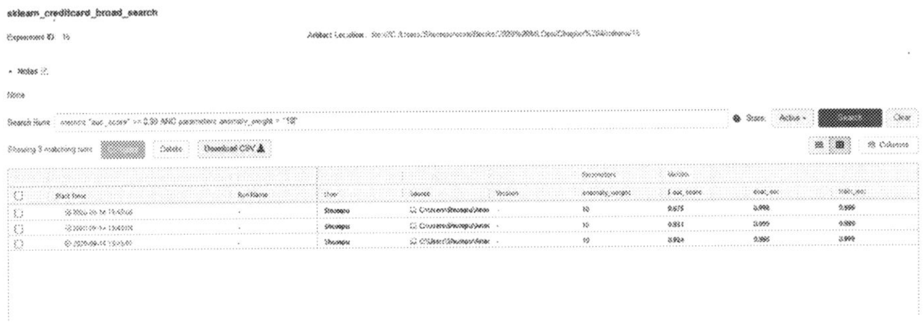

Figure 4-30. *Three runs for an anomaly weight of 10 also met your criteria for minimum AUC score*

So, three of the five folds with the anomaly weight set to 10 had an AUC score above 0.90.

Let's check 15 now. Refer to Figure 4-31.

Figure 4-31. *You can see that with an anomaly weight of 15, there seems to be two folds that had an AUC score above 0.95*

You see similar results with 15.

What if you tighten the AUC score requirement to be a minimum of 0.95? Let's check the runs for a minimum AUC of 0.95 and with an anomaly weight of 5. Refer to Figure 4-32.

Figure 4-32. *This time, you see that only one of the folds for the runs with anomaly weight set to 5 has an AUC score above 0.95*

So, it seems that only one fold reached an AUC score above 0.95 when the anomaly weight was 5.

What do the results look like for an anomaly weight of 10? Refer to Figure 4-33.

Figure 4-33. *With an anomaly weight of 10, only one run has an AUC score above 0.95*

Let's check the runs with an anomaly weight of 15. Refer to Figure 4-34.

Figure 4-34. *With an anomaly weight of 15, only one run has achieved an AUC score above 0.95. From these results, you cannot really infer which weight setting is the best, so you have to narrow the scope of your hyperparameter search. As far as you know, you could have missed the best setting, and it could be somewhere in between 5 and 10 or 10 and 15*

It seems that for an anomaly weight of 15, only one run has achieved an AUC score above 0.95. It seems that you can't look at how you can narrow the scope without looking at the rest of the AUC scores.

It appears to be the case that the best AUC scores seem to be between 5 and 15.

Alright, so what if the higher anomaly weights were more consistent in their AUC scores, and the smaller anomaly weight runs achieving the highest AUC scores were just flukes? To see how each anomaly weight setting did, first remove the query statement, and click Search again. Next, make sure that the AUC scores are in descending order. Once you're done, refer to Figure 4-35 and verify that your output looks similar.

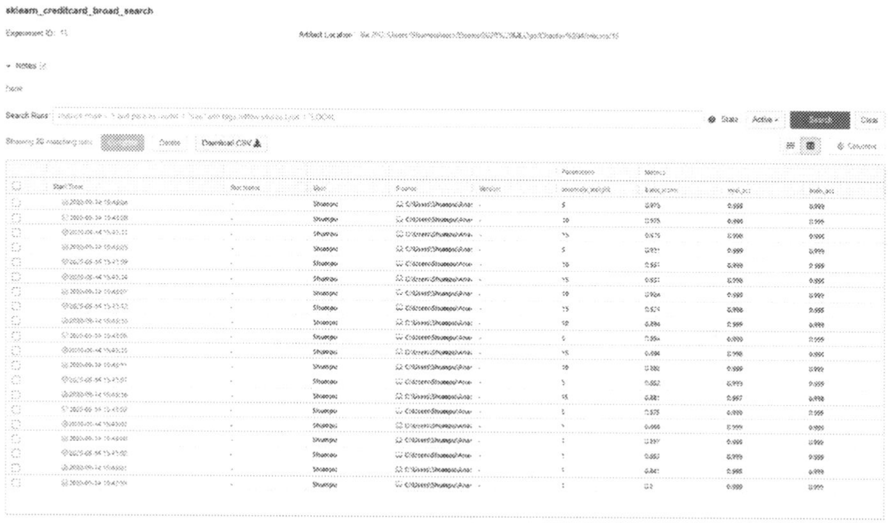

Figure 4-35. *Ordering the runs by descending AUC score*

Using the following code, let's filter over all of the values for anomaly weights and check what the AUC scores look like, replacing 1 with 5, 10, and 15.

```
parameters.anomaly_weight = "1"
```

Refer to Figure 4-36 to see the results of filtering by an anomaly weight of 1.

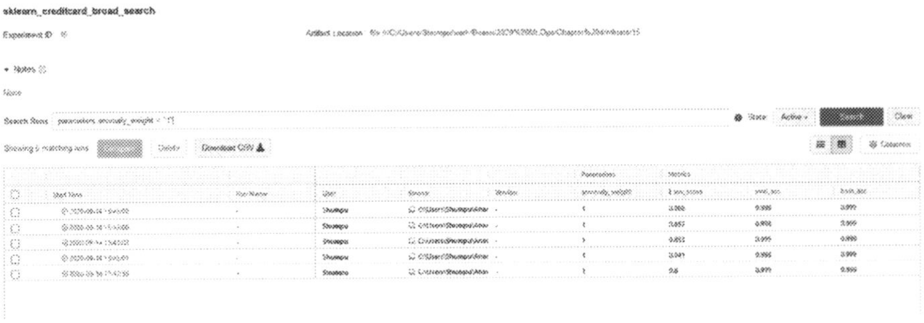

Figure 4-36. *Looking at the AUC scores of the runs with an anomaly weight of 1 in descending order*

None of the scores have gone above 0.9, so you can automatically rule out this anomaly weight setting. If you go back to your script, you can see that the average AUC was around 0.8437.

Let's look at the runs with an anomaly weight of 5. Refer to Figure 4-37.

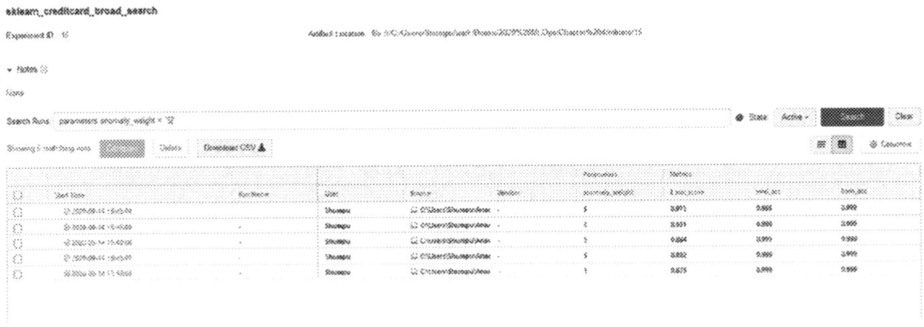

Figure 4-37. *Looking at the AUC scores of the runs with anomaly weight of 5 in descending order. You can see a noticeable increase in the average AUC score when compared to an anomaly weight of 1*

The scores have improved noticeably. If you go back to the original script's output, you can see that the average AUC score is now 0.9116.

The rest of the anomaly weights all achieved the highest AUC score of around 0.975, so the average AUC is a better metric to help you narrow the range.

Let's now look at the runs with an anomaly weight of 10. Refer to Figure 4-38.

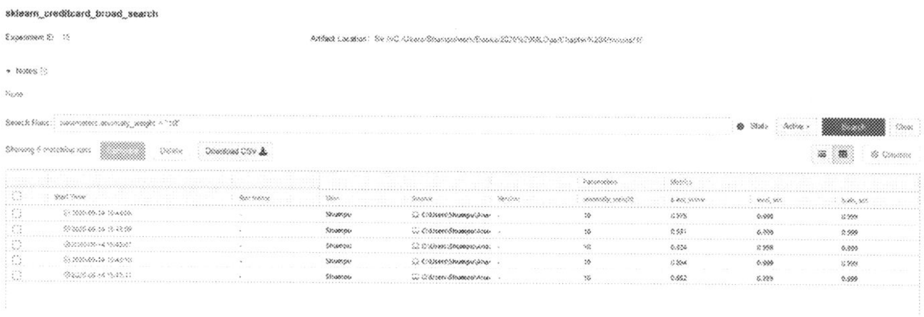

Figure 4-38. *Looking at the AUC scores of the runs with an anomaly weight of 10 in descending order. These scores seem even better*

These scores seem even better than the ones for an anomaly weight of 5. This time, the average AUC score is around 0.9215.

Finally, let's look at the scores for an anomaly weight of 15. Refer to Figure 4-39 to see the results of filtering by an anomaly weight of 15.

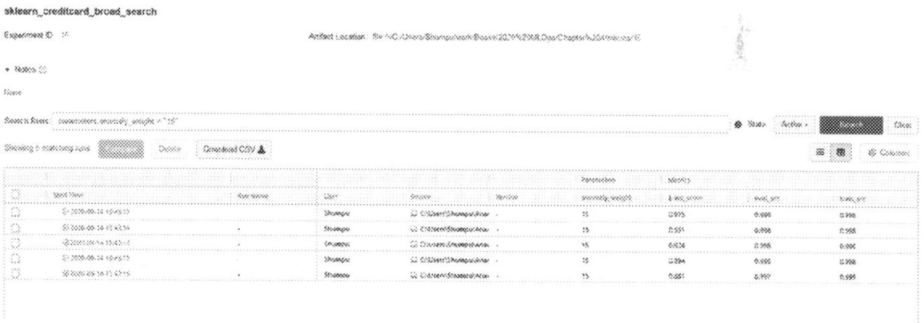

Figure 4-39. *Looking at the AUC scores of the runs with an anomaly weight of 15 in descending order. The scores are very similar, but the average is ever so slightly worse, so the true range seems to be somewhere in between 10 and 15*

The scores are very similar to each other, and indeed, the average AUC score is now 0.9212.

Based on these results, you can see that there seems to be an increase from 5 to 10, but a slight decrease from 10 to 15. From this data, the ideal range seems to be somewhere in between 10 and 15, but again, the decrease in average AUC from 10 to 15 is essentially negligible. And so, what if it's potentially beyond 15, and you started out with the wrong range to search over?

From the results of this validation experiment, it seems that you haven't found a definite range of values that you know for sure you can focus on. And so, you must expand your range even more just to see if you can get better results with higher anomaly weights.

Looking at the distribution of data and how heavily the normal points outnumber the anomalies, you can use your intuition to help guide your hyperparameter search and expand the range far more.

Now that you know this, let's try expanding the range far more.

Parameter Tuning – Guided Search

The best overall performances were achieved by anomaly weights 10 and 15, but it seems to be on an upward trend the higher up you go with the anomaly weight.

Now that you know this, let's try another validation run with a broader range of anomaly weights to try.

Go back to the cell (or copy-paste it into a new cell) in Figure 4-23 and change the anomaly weights so that they look like the following:

```
anomaly_weights = [10, 50, 100, 150, 200]
```

You should see something like Figure 4-40.

```
In [09]:  1  anomaly_weights = [10, 50, 100, 150, 200]
          2  num_folds = 5
          3  kfold = KFold(n_splits=num_folds, shuffle=True, random_state=2020)
```

Figure 4-40. *Setting a narrow range of values to search over during the second validation run*

The validation script itself should be the same, so if you simply replaced the anomaly weights in the original cell, **don't run the validation script yet!** Let's create a new experiment so that you don't clutter the original tuning experiment with these new runs.

Modify the following line in the old validation script so that it goes from

mlflow.set_experiment("*sklearn*_creditcard_broad_search")

to

mlflow.set_experiment("sklearn_creditcard_guided_search")

You should see something like Figure 4-41.

```
In [01]:  1
          2  mlflow.set_experiment("sklearn_creditcard_guided_search")
          3
          4  logs = []
          5  for f in range(len(anomaly_weights)):
          6      fold = 1
          7      accuracies = []
          8      auc_scores= []
          9      for train, test in kfold.split(x_validate, y_validate):
         10          with mlflow.start_run():
         11              weight = anomaly_weights[f]
         12              mlflow.log_param("anomaly_weight", weight)
         13  
```

Figure 4-41. *Setting a new experiment called sklearn_creditcard_ guided_search so that the results of this second validation experiment are stored separately*

Now you can run this code. Once it finishes, go back to the UI, refresh it, and select the new experiment named sklearn_creditcard_guided_ search. You should see something like Figure 4-42.

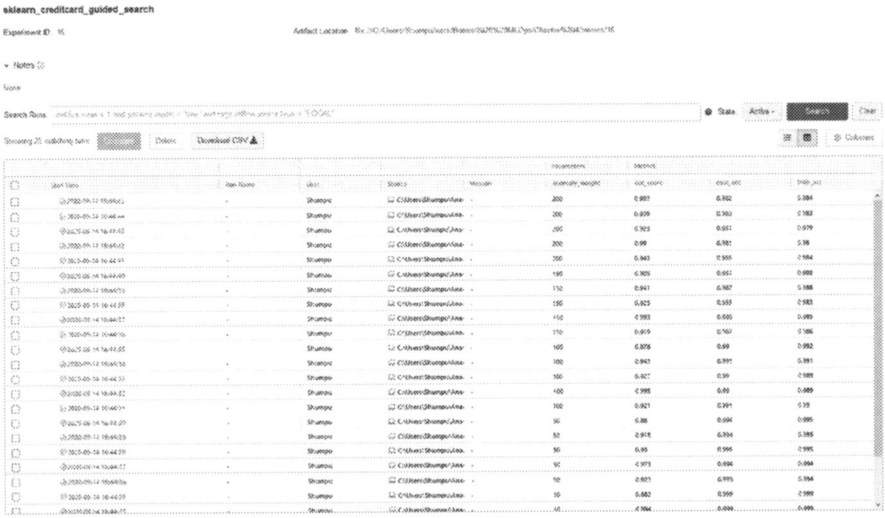

Figure 4-42. *The results of the second validation experiment*

The whole point of broadening the range of anomaly weights that you are performing the tuning experiment on is to help you understand where the best hyperparameter range may lie. You did not know this initially, so you picked a range that was too small to help you discover the best value. Now that you do know, you have expanded your search range considerably.

From the results of this experiment, you can hopefully narrow your range a lot more and repeat the experiment with a massively reduced range and arrive at the correct hyperparameter setting.

You will now filter out each of the values by each unique anomaly weight (10, 50, 100, 150, and 200) to get an idea of how the runs with that setting performed.

Make sure you're sorting AUC scores in descending order, type the following query, and search:

```
parameters.anomaly_weight = "10"
```

You should see something like Figure 4-43.

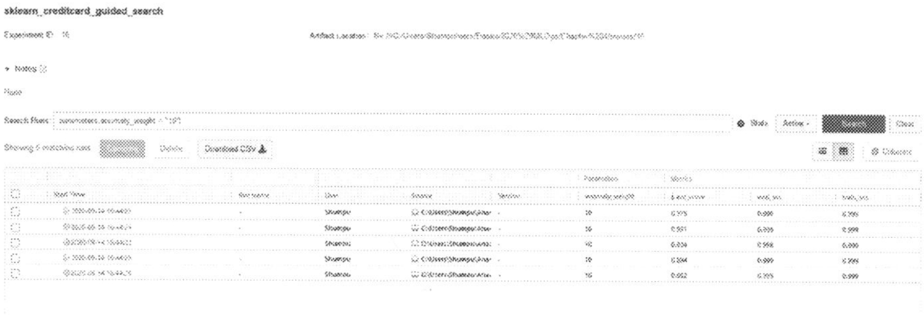

Figure 4-43. *Filtering the runs by anomaly weight of 10 and setting the AUC score to display in descending order*

The average AUC score as displayed by the validation script is around 0.9215. Of course, this is the same result as from earlier.

Let's see how the scores look for an anomaly weight of 50. Refer to Figure 4-44.

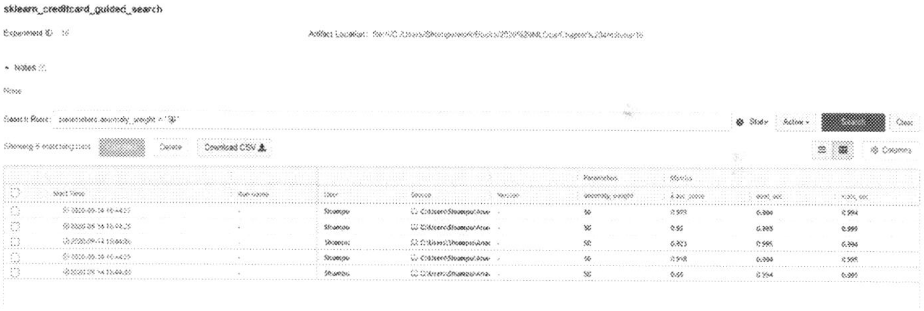

Figure 4-44. *Filtering the runs by an anomaly weight of 50 and setting the AUC score to display in descending order. It seems there's a slight difference in values*

There appears to be a minute difference in the range of AUC scores already. Looking at the script, you can see that the average AUC is around 0.9248, so there appears to be a small increase in the AUC score.

Let's keep going and check the results for the anomaly weight of 100. Refer to Figure 4-45.

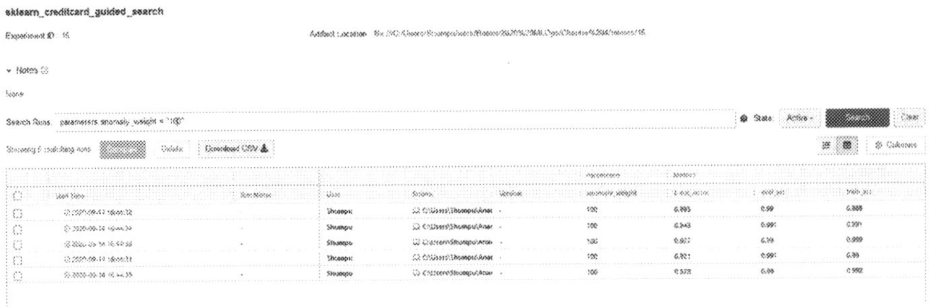

Figure 4-45. *Filtering the runs by an anomaly weight of 100 and setting the AUC score to display in descending order*

The average this time appears to be 0.9327. Despite the massive increase in weight, the average AUC score did not go up that high. However, notice that the first result with an AUC score of 0.995 has appeared. The best AUC score up until the anomaly weight of 50 was 0.975, but this anomaly weight setting has broken past that.

Let's keep going and see if it increases with an anomaly weight setting of 150. Refer to Figure 4-46A.

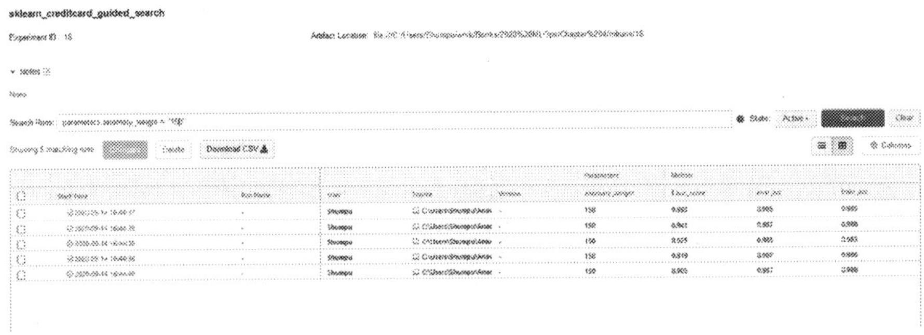

Figure 4-46A. *Filtering the runs by an anomaly weight of 150 and setting the AUC score to display in descending order*

The AUC scores overall seem to be a bit higher. Indeed, the average AUC score is now 0.9365, so there was an increase. Finally, let's check the AUC scores for an anomaly weight setting of 200. Refer to Figure 4-46B.

Figure 4-46B. *Filtering the runs by an anomaly weight of 200 and setting the AUC score to display in descending order*

The new average AUC now is 0.9396, so this anomaly weight setting seems even better.

In fact, you still weren't able to come to a conclusion about an optimal range, since the AUC scores keep increasing as you set higher anomaly weights.

So, from this, you know that the best hyperparameter setting is somewhere above 200. You simply shift the range of the scope to start at 200 and search over a slightly different area, and once you have found a good range of values to search over (eventually the AUC scores will start trending down as you increase the anomaly weight), you can narrow the focus and start searching again.

After a certain amount of precision with the parameter value, you start to see diminishing returns where the added effort only produces negligible improvements in performance, but you will likely encounter this as you start getting deeper into the decimal values.

Hopefully now you understand more about how you can integrate MLFlow into the model training, testing, and validation pipeline using scikit-learn. You also looked at how to use the UI for basic comparisons, along with how you might perform hyperparameter tuning more easily using MLFlow.

A quick note to make is that if you'd like to perform more complicated searches over multiple metrics or parameters, MLFlow provides functionality through the API to let you do so via SQL searches within the code, letting you order by multiple columns, for example.

MLFlow also provides support for logging metrics, parameters, artifacts, and even models for other frameworks in their documentation. We will now take a look at how to integrate MLFlow with TensorFlow 2.0+/ Keras, PyTorch, and PySpark.

MLFlow and Other Frameworks

MLFlow with TensorFlow 2.0 (Keras)

MLFlow provides easy integration with TensorFlow 2.0+ (any version of TensorFlow 2.0 and above). To see how, let's go over a very basic example of a handwritten digit classifier model on the MNIST dataset. We will be using the built-in Keras module to keep things simple for demonstration purposes. MLFlow supports TensorFlow 1.12 at a minimum, so this code should run as long as you have at least TensorFlow 1.12.

We will assume a basic level of familiarity with TensorFlow 2, so we won't go into much depth about what the functions, model layers, optimizers, and loss functions mean.

Before we begin, here are the versions of TensorFlow, CUDA, and CuDNN that we used. Keep in mind that we ran this using the GPU version of TensorFlow (the package is called tensorflow-gpu), although you should be able to run this without a GPU at the cost of it taking longer:

- **TensorFlow** (GPU version) – 2.3.0

- **CUDA** – 10.1

- **CuDNN** – v7.6.5.32 for CUDA 10.1

- **Sklearn** – 0.22.2.post1

- **MLFlow** – 1.10.0

Data Processing

Here is the code to import the necessary modules and print out their versions:

```
import tensorflow as tf
from tensorflow.keras.models import Sequential
from tensorflow.keras.layers import Dense, Conv2D, Flatten
from tensorflow.keras.datasets import mnist

import numpy as np

import matplotlib
import matplotlib.pyplot as plt

import sklearn
from sklearn.metrics import roc_auc_score

import mlflow
import mlflow.tensorflow

print("TensorFlow: {}".format(tf.__version__))
print("Scikit-Learn: {}".format(sklearn.__version__))
print("Numpy: {}".format(np.__version__))
print("MLFlow: {}".format(mlflow.__version__))
print("Matplotlib: {}".format(matplotlib.__version__))
```

You should see something like Figure 4-47.

```
In [13]:    1  import tensorflow as tf
            2  from tensorflow.keras.models import Sequential
            3  from tensorflow.keras.layers import Dense, Conv2D, Flatten
            4  from tensorflow.keras.datasets import mnist
            5
            6  import numpy as np
            7
            8  import matplotlib
            9  import matplotlib.pyplot as plt
           10
           11  import sklearn
           12  from sklearn.metrics import roc_auc_score
           13
           14  import mlflow
           15  import mlflow.tensorflow
           16
           17  print("TensorFlow: {}".format(tf.__version__))
           18  print("Scikit-Learn: {}".format(sklearn.__version__))
           19  print("Numpy: {}".format(np.__version__))
           20  print("MLFlow: {}".format(mlflow.__version__))
           21  print("Matplotlib: {}".format(matplotlib.__version__))
           22

        TensorFlow: 2.3.0
        Scikit-Learn: 0.22.2.post1
        Numpy: 1.18.5
        MLFlow: 1.10.0
        Matplotlib: 3.2.1
```

Figure 4-47. *Importing the necessary modules and printing their versions*

Let's now load the data:

```
(x_train, y_train), (x_test, y_test) = mnist.load_data()
```

Keras, and by extension TensorFlow, provides the MNIST handwritten digit dataset for you, so all you need to do to load the data is call the function, like in Figure 4-48.

Refer to Figure 4-48 to see the code in a cell.

```
In [6]:    1  (x_train, y_train), (x_test, y_test) = mnist.load_data()
```

Figure 4-48. *Defining x_train, y_train, x_test, and y_test*

You can even see what one of these images looks like. Run the following:

```
plt.imshow(x_train[0], cmap='gray'), print("Class: ", y_train[0])
```

You should see something like Figure 4-49.

Figure 4-49. *Looking at what one of the data samples looks like using matplotlib. You also printed out the class label associated with this sample, which was 5*

Also notice that you printed out the class label associated with this specific image. The labels are all integers between 0 and 9, each associated with an image that shows a handwritten digit from 0 to 9.

Since 2D convolutional layers in TensorFlow/Keras expect four dimensions in the format of (m, h, w, c) where m stands for the number of samples in the dataset, h and w stand for the height and width, respectively, and c stands for the number of channels (three if it's an RGB color image for example), you need to reshape your data so that it conforms to these specifications. Your images are all black and white, so they technically have a channel of one. And so, you must reshape them like so:

```
x_train = x_train.reshape(x_train.shape[0], x_train.shape[1],
x_train.shape[2], 1)
x_test = x_test.reshape(x_test.shape[0], x_test.shape[1],
x_test.shape[2], 1)

y_train = tf.keras.utils.to_categorical(y_train)
y_test = tf.keras.utils.to_categorical(y_test)
```

Refer to Figure 4-50 to see that code in a cell.

```
In [8]:   1  x_train = x_train.reshape(x_train.shape[0], x_train.shape[1], x_train.shape[2], 1)
          2  x_test = x_test.reshape(x_test.shape[0], x_test.shape[1], x_test.shape[2], 1)
          3
          4  y_train = tf.keras.utils.to_categorical(y_train)
          5  y_test = tf.keras.utils.to_categorical(y_test)
```

Figure 4-50. *Reshaping the data to include one channel, conforming with the specifications of the convolutional layers. Additionally, the y sets are being converted to one-hot encoded formats*

You converted the y sets by calling a function called
to_categorical(). This converts each sample from an integer value of
say 2 or 4 corresponding to the digit represented by the x samples into a
one-hot encoded vector.

Samples in this format are now 0 vectors with a num_classes number
of digits. In other words, these vectors all have a length matching the total
number of classes. Whatever value the label was is now the index of the
value 1. And so, if the label is 1, the value at the index of 1 in this vector will
be one, and everything else is a 0.

This may be a little confusing, so refer to Figure 4-51 to see what the
one-hot encoded label looks like for a digit representing 5.

```
In [9]:   1  y_train[0]
Out[9]:   array([0., 0., 0., 0., 0., 1., 0., 0., 0., 0.], dtype=float32)
```

Figure 4-51. *The new output of the one-hot encoded label representing a value of 5. Notice that the value at index 5 is now 1*

As you can see, the index of the 1 is 5, corresponding to the first
x_train example you looked at earlier, which was the digit 5.

Now, let's print out the shapes:

```
print("Shapes")
print("x_train: {}\ny_train: {}".format(x_train.shape,
y_train.shape))
print("x_test: {}\ny_test: {}".format(x_test.shape,
y_test.shape))
```

You should now see something like Figure 4-52.

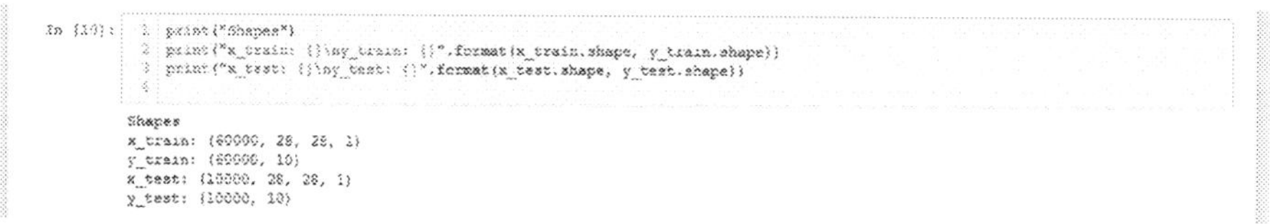

```
In [19]:    1  print("Shapes")
            2  print("x_train: {}\ny_train: {}".format(x_train.shape, y_train.shape))
            3  print("x_test: {}\ny_test: {}".format(x_test.shape, y_test.shape))
            4

Shapes
x_train: (60000, 28, 28, 1)
y_train: (60000, 10)
x_test: (10000, 28, 28, 1)
y_test: (10000, 10)
```

Figure 4-52. *Printing the output shapes of the processed data*

MLFlow Run – Training and Evaluating

Let's move on to the creation of your model. You will be using the
Sequential method of model creation. The model will be quite simple,
consisting of a couple 2D convolutional layers that feed into three dense
layers. Run the following:

```
model = Sequential()

model.add(Conv2D(filters=16, kernel_size=3, strides=2,
padding='same', input_shape=(28, 28, 1), activation="relu"))
model.add(Conv2D(filters=8, kernel_size=3, strides=2,
padding='same', input_shape=(28, 28, 1), activation="relu"))
model.add(Flatten())
model.add(Dense(30, activation="relu"))
model.add(Dense(20, activation="relu"))
model.add(Dense(10, activation="softmax"))

model.summary()
```

You should see something like Figure 4-53.

```
In [6]:    1  model = Sequential()
           2
           3  model.add(Conv2D(filters=16, kernel_size=3, strides=2, padding='same', input_shape=(28, 28, 1), activation="relu"))
           4  model.add(Conv2D(filters=8, kernel_size=3, strides=2, padding='same', input_shape=(28, 28, 1), activation="relu"))
           5  model.add(Flatten())
           6  model.add(Dense(30, activation="relu"))
           7  model.add(Dense(20, activation="relu"))
           8  model.add(Dense(10, activation="softmax"))
           9
          10  model.summary()

Model: "sequential"

Layer (type)                 Output Shape              Param #
=================================================================
conv2d (Conv2D)              (None, 14, 14, 16)        160
_____
conv2d_1 (Conv2D)            (None, 7, 7, 8)           1160
_____
flatten (Flatten)            (None, 392)               0
_____
dense (Dense)                (None, 30)                11790
_____
dense_1 (Dense)              (None, 20)                620
_____
dense_2 (Dense)              (None, 10)                210
=================================================================
Total params: 13,940
Trainable params: 13,940
Non-trainable params: 0
_____
```

Figure 4-53. *Creating the model and outputting a summary of the model's architecture*

Let's now compile your model using the Adam optimizer and categorical cross-entropy for your loss. For your metric, you will only be using accuracy. Run the following:

```
model.compile(optimizer="Adam",
loss="categorical_crossentropy", metrics=["accuracy"])
```

You should see something like Figure 4-54.

```
In [15]:   1  model.compile(optimizer="Adam", loss="categorical_crossentropy", metrics=["accuracy"])
           4
```

Figure 4-54. *Compiling your model, setting the optimizer to Adam optimizer, setting the loss to categorical cross-entropy, and setting the metric to be accuracy*

Now you get to the part where you tell MLFlow to log this run. You want all of the metrics to be logged to the same run, so you must tell MLFlow specifically to run a block of code in the context of the same run. To do so, you once again block your code using the following line:

```
with mlflow.start_run():
```

With that, run the following to set the experiment name, train the model, get the evaluation metrics you need, and log them all to MLFlow:

```
mlflow.set_experiment("TF_Keras_MNIST")

with mlflow.start_run():

    mlflow.tensorflow.autolog()

    model.fit(x=x_train, y=y_train, batch_size=256, epochs=10)
    preds = model.predict(x_test)
    preds = np.round(preds)

    eval_acc = model.evaluate(x_test, y_test)[1]
    auc_score = roc_auc_score(y_test, preds)

    print("eval_acc: ", eval_acc)
    print("auc_score: ", auc_score)

    mlflow.tensorflow.mlflow.log_metric("eval_acc", eval_acc)
    mlflow.tensorflow.mlflow.log_metric("auc_score", auc_score)

mlflow.end_run()
```

Refer to Figure 4-55 to see the output. Ignore the warning messages. They don't hinder the training process or the performance of the model.

```
In [9]:   1  mlflow.set_experiment("TF_Keras_MNIST")
          2
          3  with mlflow.start_run():
          4
          5      mlflow.tensorflow.autolog()
          6
          7      model.fit(x=x_train, y=y_train, batch_size=256, epochs=10)
          8      preds = model.predict(x_test)
          9      preds = np.round(preds)
         10
         11      eval_acc = model.evaluate(x_test, y_test)[1]
         12      auc_score = roc_auc_score(y_test, preds)
         13
         14      print("eval_acc: ", eval_acc)
         15      print("auc_score: ", auc_score)
         16
         17      mlflow.tensorflow.mlflow.log_metric("eval_acc", eval_acc)
         18      mlflow.tensorflow.mlflow.log_metric("auc_score", auc_score)
         19
         20  mlflow.end_run()
```

```
INFO: 'TF_Keras_MNIST' does not exist. Creating a new experiment
Epoch 1/10
   1/235 [..............................] - ETA: 0s - loss: 9.2293 - accuracy: 0.0859WARNING:tensorflow:From
C:\Users\Shumpu\Anaconda2\envs\p36\lib\site-packages\tensorflow\python\ops\summary_ops_v2.py:1277: stop (from
tensorflow.python.eager.profiler) is deprecated and will be removed after 2020-07-01.
Instructions for updating:
use `tf.profiler.experimental.stop` instead.
   2/235 [..............................] - ETA: 34s - loss: 8.2826 - accuracy: 0.0996WARNING:tensorflow:Callba
cks method `on_train_batch_begin` is slow compared to the batch time (batch time: 0.0050s vs `on_train_batch_b
egin` time: 0.8430s). Check your callbacks.
WARNING:tensorflow:Callbacks method `on_train_batch_end` is slow compared to the batch time (batch time: 0.005
0s vs `on_train_batch_end` time: 0.2410s). Check your callbacks.
235/235 [==============================] - 2s 8ms/step - loss: 1.1566 - accuracy: 0.6513
Epoch 2/10
235/235 [==============================] - 1s 6ms/step - loss: 0.2496 - accuracy: 0.9252
Epoch 3/10
235/235 [==============================] - 1s 6ms/step - loss: 0.1542 - accuracy: 0.9544
Epoch 4/10
235/235 [==============================] - 1s 6ms/step - loss: 0.1139 - accuracy: 0.9664
Epoch 5/10
235/235 [==============================] - 1s 6ms/step - loss: 0.0903 - accuracy: 0.9730
Epoch 6/10
235/235 [==============================] - 1s 6ms/step - loss: 0.0756 - accuracy: 0.9771
Epoch 7/10
235/235 [==============================] - 1s 6ms/step - loss: 0.0699 - accuracy: 0.9803
Epoch 8/10
235/235 [==============================] - 1s 6ms/step - loss: 0.0575 - accuracy: 0.9826
Epoch 9/10
235/235 [==============================] - 1s 6ms/step - loss: 0.0594 - accuracy: 0.9839
Epoch 10/10
235/235 [==============================] - 1s 6ms/step - loss: 0.0444 - accuracy: 0.9860
313/313 [==============================] - 1s 4ms/step - loss: 0.0789 - accuracy: 0.9767
eval_acc:  0.9767000079154968
auc_score:  0.986283036190461
```

Figure 4-55. *Output of the MLFlow run and the training process. You can also see that the metrics you calculated have been updated*

Another new line of code is the following:

```
mlflow.keras.autolog()
```

This basically tells MLFlow to log all the parameters and metrics associated with the particular TensorFlow/Keras model. As you will see shortly, MLFlow will log the hyperparameters, model metrics listed in the compile() function, and even the model itself once the training has finished.

MLFlow UI – Checking Your Run

With that, let's now open the MLFlow UI and check your run in MLFlow. Make sure your terminal or command prompt is in the same directory where the mlruns are stored. Usually, MLFlow saves all these runs in the same directory of the Jupyter notebook.

Now that you've opened the UI, you should see something like Figure 4-56.

Figure 4-56. *The MLFlow UI after running the TensorFlow experiment. Notice that there is a new experiment titled TF_Keras_ MNIST*

Click the tab called TF_Keras_MNIST to see the results of the experiment you just logged. You should see something like Figure 4-57.

Figure 4-57. *Opening the experiment titled TF_Keras_MNIST. You can see that it successfully logged a run*

As you can see, your run was just successfully logged. Next, click it to explore all of the parameters, metrics, and artifacts that MLFlow logged.

You should see something like Figure 4-58.

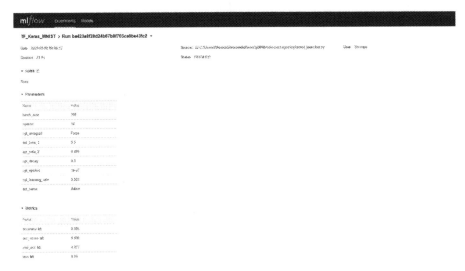

Figure 4-58. *Looking at the specific run logged in the experiment. As you can see, all the parameters and metrics were logged, even the one you specified. It also shows you the duration and the status of the run, so now you know how long it took to train the model as well as whether or not it completed*

MLFlow saved all of the hyperparameters used when creating the model. This could be very useful for hyperparameter tuning on a validation set, for example, where you are trying to tune many hyperparameters at once. For example, you can definitely tune batch_size, epochs, or something related to the Adam optimizer like opt_learning_rate, opt_beta_1, or opt_beta_2.

As you can see in Figure 4-58, MLFlow saved the model metrics for accuracy and loss as calculated during the training process. In addition, MLFlow also saved the metrics that you defined.

Scroll down to artifacts and click model and then data. You should see something like Figure 4-59.

Figure 4-59. *Upon closer inspection of the artifacts, it seems MLFlow has also logged the model itself*

Here, you can see that MLFlow also saved the model after the training process finished. In fact, let's briefly look at how you can load this model. Make sure you go to the top and copy the run ID before doing this.

Loading an MLFlow Model

With the run ID copied, head on over to the notebook and create a new cell. Run the following code, but replace the run ID with yours:

```
loaded_model =
mlflow.keras.load_model("runs:/YOUR_RUN_ID/model")
```

Your code should look similar to Figure 4-60. Our run was
ba423a8f28d24b67b8f703ca6be43fc2, so that's what we replaced
YOUR_RUN_ID with.

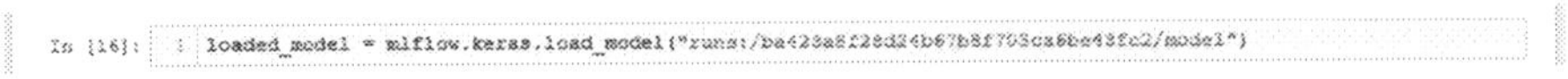

```
In [16]:    1  loaded_model = mlflow.keras.load_model("runs:/ba423a8f28d24b67b8f703ca6be43fc2/model")
```

Figure 4-60. *Loading a logged model using a specific run. Notice that
we are doing mlflow.keras. This is because the model is technically a
Keras model*

You'll notice that we did `mlflow.keras` instead of `mlflow.tensorflow`.
This is because this model is technically a Keras model, and so it conforms
to the specific `load_model()` code in the `mlflow.keras` module.

Run the following code to quickly calculate the same evaluation
metrics that you logged earlier:

```
eval_loss, eval_acc = loaded_model.evaluate(x_test, y_test)

preds = loaded_model.predict(x_test)
preds = np.round(preds)

eval_auc = roc_auc_score(y_test, preds)

print("Eval Loss:", eval_loss)
print("Eval Acc:", eval_acc)
print("Eval AUC:", eval_auc)
```

This just ensures that the model is the same and demonstrates that
you can use the model to make predictions. Refer to Figure 4-61 to see the
output.

```
In [21]:  1  eval_loss, eval_acc = loaded_model.evaluate(x_test, y_test)
          2
          3  preds = loaded_model.predict(x_test)
          4  preds = np.round(preds)
          5
          6  eval_auc = roc_auc_score(y_test, preds)
          7
          8  print("Eval Loss:", eval_loss)
          9  print("Eval Acc:", eval_acc)
         10  print("Eval AUC:", eval_auc)

313/313 [==============================] - 1s 4ms/step - loss: 0.0789 - accuracy: 0.9767
Eval Loss: 0.07890588045120239
Eval Acc: 0.9767000079154968
Eval AUC: 0.986283036190481
```

Figure 4-61. *The output of the code block printing out the loss,
accuracy, and AUC score when the model was evaluated on the test
set. These three values match the corresponding values from the
output of the run earlier*

As you can see, this output matches the values from the output of
the run earlier. Additionally, this model is also functional and can make
predictions.

And with that, you now know how to integrate MLFlow into your
TensorFlow 2.0+ experiments. Again, MLFlow supports TensorFlow 1.12+,
which also contains the Keras submodule. This means that you should be
able to follow the same convention to log tf.keras module code as long as
you have TensorFlow 1.12+.

In practice, you are likely to have functions to build and compile the
model, functions to train the model, and functions to evaluate and perhaps
even validate the model. Just be sure to call all of them in the block `with`
`mlflow.start_run():` so that MLFlow knows all of this is happening
within the same run.

Next, let's look at how to integrate MLFlow with PyTorch.

MLFlow with PyTorch

MLFlow also provides integration with PyTorch. While the process isn't as
easy as with Keras or TensorFlow, integrating MLFlow into your existing
PyTorch code is quite simple. To see how to do so, we will be exploring a
simple convolutional neural network applied to the MNIST dataset once
again.

Before we begin, here are the versions of the modules we are using, including CUDA and CuDNN:

- **Torch** - 1.6.0

- **Torchvision** – 0.7.0

- **CUDA** – 10.1

- **CuDNN** – v7.6.5.32 for CUDA 10.1

- **Sklearn** – 0.22.2.post1

- **MLFlow** – 1.10.0

- **numpy** – 1.18.5

Data Processing

Let's get started. Here's the code to import the necessary modules, print out their versions, and set the device that PyTorch will use:

```python
import torch
import torch.nn as nn
from torch.utils import data

import torchvision
import torchvision.datasets

import sklearn
from sklearn.metrics import roc_auc_score, accuracy_score

import numpy as np

import mlflow
import mlflow.pytorch

device = torch.device("cuda:0" if torch.cuda.is_available()
else "cpu")
```

```
print("PyTorch: {}".format(torch.__version__))
print("torchvision: {}".format(torchvision.__version__))
print("sklearn: {}".format(sklearn.__version__))
print("MLFlow: {}".format(mlflow.__version__))
print("Numpy: {}".format(np.__version__))
print("Device: ", device)
```

Refer to Figure 4-62 to see the output.

Figure 4-62. *Importing the necessary modules and printing the versions of the modules*

The line of code

```
device = torch.device("cuda:0" if torch.cuda.is_available()
else "cpu")
```

tells PyTorch which device to run the code on. If there is a GPU that CUDA can connect to, it will use that instead. Otherwise, it will run everything on the CPU. In our case, we have CUDA set up with our GPU, so Torch displays "cuda:0" as seen in Figure 4-62.

Next, you will define some basic hyperparameters:

```
batch_size = 256
num_classes = 10
learning_rate = 0.001
```

Refer to Figure 4-63 to see them in a cell.

Figure 4-63. *Setting the hyperparameters relevant to the training of the model*

Next, you will load in the MNIST dataset. Like Keras and TensorFlow, PyTorch also provides example datasets. In this case, you are loading MNIST:

```
train_set = torchvision.datasets.MNIST(root='./data',
train=True, download=True, transform=None)
test_set = torchvision.datasets.MNIST(root='./data',
train=False, download=True, transform=None)
```

Refer to Figure 4-64 to see this code in a cell.

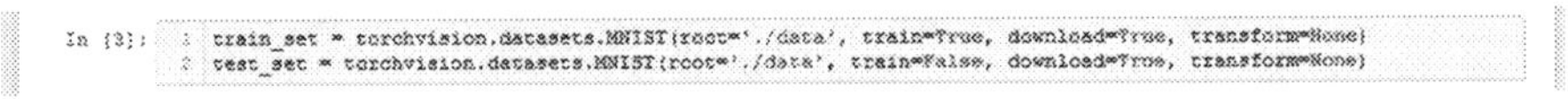

Figure 4-64. *Defining the training and testing sets by loading the data from PyTorch*

You will now define your x_train, y_train, x_test, and y_test datasets:

```
x_train, y_train = train_set.data, train_set.targets
x_test, y_test = test_set.data, test_set.targets
```

Refer to Figure 4-65.

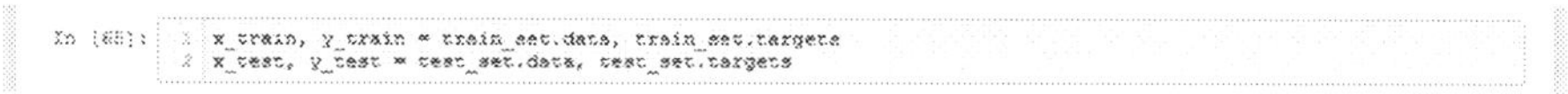

```
In [68]:    1  x_train, y_train = train_set.data, train_set.targets
            2  x_test, y_test = test_set.data, test_set.targets
```

Figure 4-65. *Creating your x_train, y_train, x_test, and y_test data sets from the training and testing sets*

In PyTorch, you want the data to be channels first. In other words, the format of the data should be (m, c, h, w), where *m* stands for the number of samples, *c* stands for the number of channels, *h* stands for the height of the samples, and *w* stands for the width of the samples.

Notice that this is the "opposite" format of how Keras and TensorFlow do it by default, which is channels last. In Keras and TensorFlow, you can also do channels first, but you must specify that you are doing it this way.

Let's reshape your x-sets:

```
x_train, y_train = train_set.data, train_set.targets
x_test, y_test = test_set.data, test_set.targets
```

Refer to Figure 4-66 to see this code in a cell.

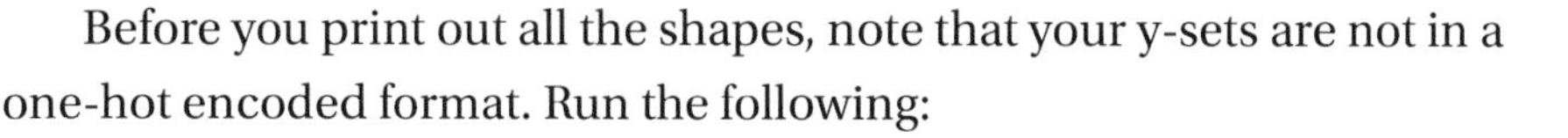

```
In [5]:    1  x_train = x_train.reshape(x_train.shape[0], 1, x_train.shape[1], x_train.shape[2])
           2  x_test = x_test.reshape(x_test.shape[0], 1, x_test.shape[1], x_test.shape[2])
           3
```

Figure 4-66. *Reshaping the x-sets so the data is encoded in a channels-first format*

Before you print out all the shapes, note that your y-sets are not in a one-hot encoded format. Run the following:

```
y_train[0]
```

Refer to Figure 4-67.

```
In [6]:    1  y_train[0]
Out[6]:    tensor(5)
```

Figure 4-67. *The output of the first sample in the y_train set. Note that the numbers are not in a one-hot encoded format*

Notice that this outputs a number, not a vector. You must convert your y-sets into a one-hot encoded format. However, there isn't a handy function like keras.utils.to_categorical() you can just call, so you will define one:

```
def to_one_hot(num_classes, labels):
    one_hot = torch.zeros(([labels.shape[0], num_classes]))
    for f in range(len(labels)):
        one_hot[f][labels[f]] = 1

    return one_hot
```

That being said, you can always call keras.utils.to_categorical(): and type-cast the resulting output to a PyTorch tensor.

Refer to Figure 4-68 to see this in a cell.

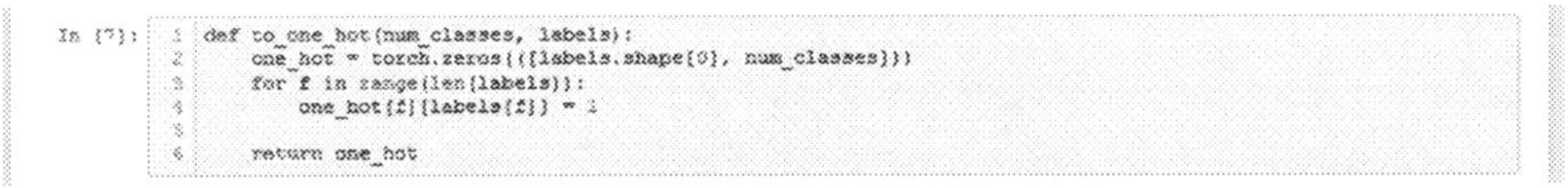

Figure 4-68. *A custom function that converts the input called "labels," given the number of classes, into a one-hot encoded format and returns it*

Now let's convert your y-sets to be in a one-hot encoded format:

```
y_train = to_one_hot(num_classes, y_train)
y_test = to_one_hot(num_classes, y_test)
```

Refer to Figure 4-69 to see this code in a cell.

Figure 4-69. *Converting your y-sets into a one-hot encoded format using your custom function*

Let's check what y_train looks like now:

y_train[0]

Refer to Figure 4-70.

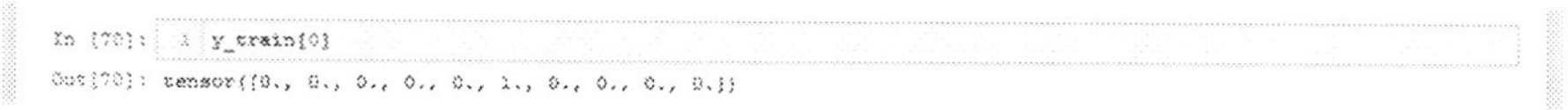

```
In [70]:  1  y_train[0]
Out[70]:  tensor([0., 0., 0., 0., 0., 1., 0., 0., 0., 0.])
```

Figure 4-70. *Checking the output of the first sample in y_train, you now see that the tensor has been converted into a one-hot encoded format*

As you can see, it is now in a one-hot encoded format. Now you can proceed to checking the shapes of your data sets:

```
print("Shapes")
print("x_train: {}\ny_train: {}".format(x_train.shape,
y_train.shape))
print("x_test: {}\ny_test: {}".format(x_test.shape,
y_test.shape))
```

You should see something like Figure 4-71.

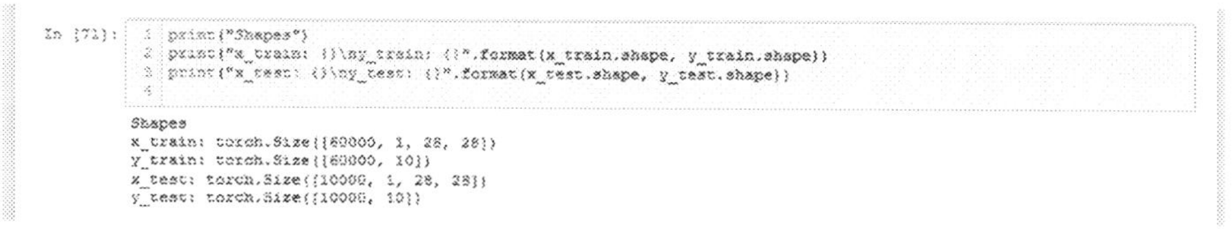

```
In [71]:  1  print("Shapes")
          2  print("x_train: {}\ny_train: {}".format(x_train.shape, y_train.shape))
          3  print("x_test: {}\ny_test: {}".format(x_test.shape, y_test.shape))
          4

Shapes
x_train: torch.Size([60000, 1, 28, 28])
y_train: torch.Size([60000, 10])
x_test: torch.Size([10000, 1, 28, 28])
y_test: torch.Size([10000, 10])
```

Figure 4-71. *Printing the shapes of your training and testing sets. As you can see, the x-sets are in a channels-first format, and the y-sets are in a one-hot encoded format*

MLFlow Run – Training and Evaluating

Now, let's define your model. A popular convention in PyTorch is to define the model as a class since it allows you to much more easily use the GPU while training. Instead of passing in every layer to the GPU, you can just send in the model object directly.

Run the following code to define your model:

```
class model(nn.Module):
    def __init__(self):
        super(model, self).__init__()

        # IN 1x28x28 OUT 16x14x14
        self.conv1 = nn.Conv2d(in_channels=1, out_channels=16,
        kernel_size=3, stride=2, padding=1, dilation=1)
        # IN 16x14x14 OUT 32x6x6
        self.conv2 = nn.Conv2d(in_channels=16, out_channels=32,
        kernel_size=3, stride=2, padding=0, dilation=1)
        # IN 32x6x6 OUT 64x2x2
        self.conv3 = nn.Conv2d(in_channels=32, out_channels=64,
        kernel_size=3, stride=2, padding=0, dilation=1)
        # IN 64x2x2 OUT 256
        self.flat1 = nn.Flatten()
        self.dense1 = nn.Linear(in_features=256,
        out_features=128)
        self.dense2 = nn.Linear(in_features=128,
        out_features=64)
        self.dense3 = nn.Linear(in_features=64,
        out_features=10)

    def forward(self, x):
        x = self.conv1(x)
        x = nn.ReLU()(x)
```

```
x = self.conv2(x)
x = nn.ReLU()(x)
x = self.conv3(x)
x = nn.ReLU()(x)
x = self.flat1(x)
x = self.dense1(x)
x = nn.ReLU()(x)
x = self.dense2(x)
x = nn.ReLU()(x)
x = self.dense3(x)
x = nn.Softmax()(x)
return x
```

Refer to Figure 4-72.

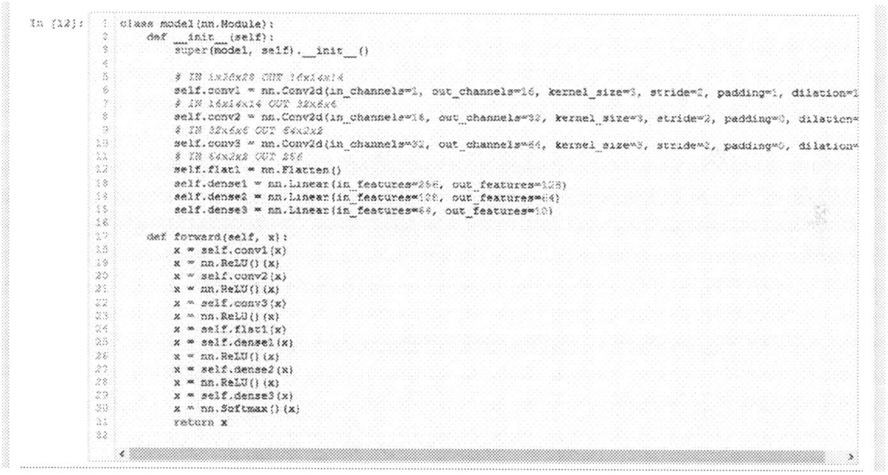

Figure 4-72. *Defining the model's architecture as a class*

Next, let's send your model to the device, define and initialize an instance of Adam optimizer with the learning rate you set earlier, and set your loss function:

```
model = model().to(device)
optimizer = torch.optim.Adam(model.parameters(),
lr=learning_rate)
criterion = nn.BCELoss()
```

Refer to Figure 4-73.

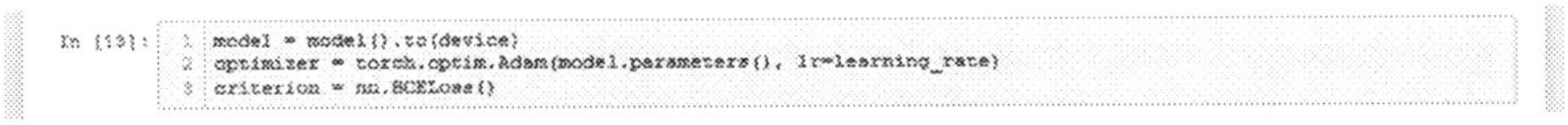

```
In [13]:  1  model = model().to(device)
          2  optimizer = torch.optim.Adam(model.parameters(), lr=learning_rate)
          3  criterion = nn.BCELoss()
```

Figure 4-73. *Sending the model object to the device, defining your optimizer, and initializing the loss function*

Next, you will define a data loader using functionality provided by PyTorch to take care of batching your data set:

```
dataset = data.TensorDataset(x_train,y_train)
train_loader = data.DataLoader(dataset, batch_size=batch_size)
```

Refer to Figure 4-74.

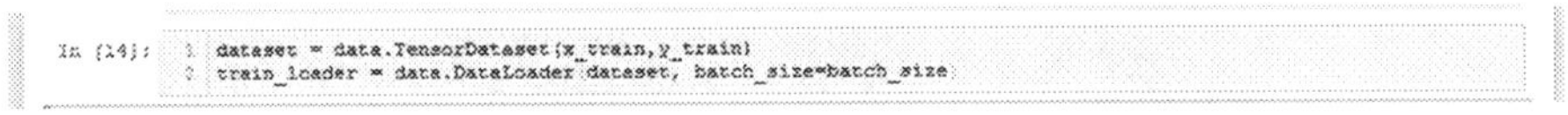

```
In [14]:  1  dataset = data.TensorDataset(x_train,y_train)
          2  train_loader = data.DataLoader(dataset, batch_size=batch_size)
```

Figure 4-74. *Creating a data loader object out of your data set. With this functionality, PyTorch will batch your data set for you, allowing you to pass in a minibatch at a time in your training loop. This essentially is what the TensorFlow 2/Keras .fit() function does, but it's all abstracted for you*

As you can see, this is much simpler than having to make an intricate loop to batch and pass in data yourself.

Finally, let's define the training loop:

```
num_epochs = 5
for f in range(num_epochs):
    for batch_num, minibatch in enumerate(train_loader):
        minibatch_x, minibatch_y = minibatch[0], minibatch[1]

        output = model.forward(torch.Tensor
        (minibatch_x.float()).cuda())
        loss = criterion(output, torch.Tensor
        (minibatch_y.float()).cuda())

        optimizer.zero_grad()
        loss.backward()
        optimizer.step()

        print(f"Epoch {f} Batch_Num {batch_num} Loss {loss}")
```

This can take at least a couple minutes depending on your GPU, and even longer if you're using a CPU. Feel free to lower the number of epochs if you'd like to decrease total training time.

You should see an output like Figure 4-75.

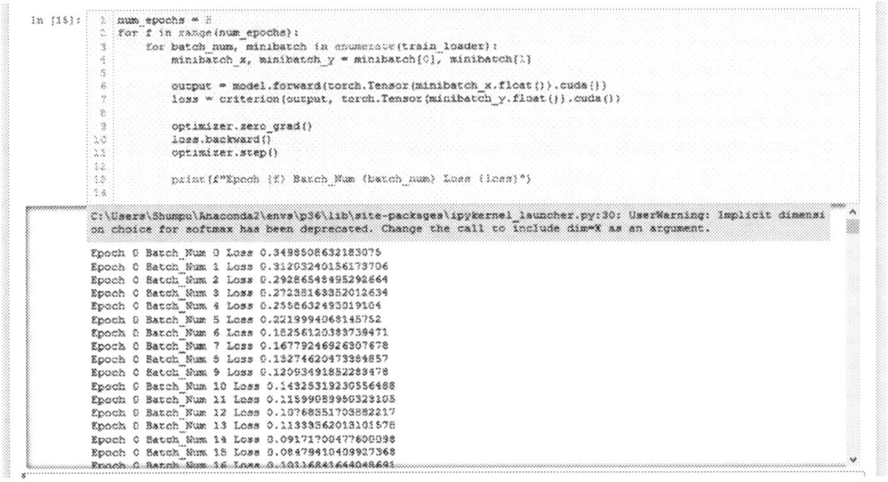

Figure 4-75. *Output of your training loop. Feel free to reduce the number of epochs to save on training time, but this could potentially hinder the model's performance*

Now, let's start an MLFlow run, calculate the metrics you want, and log everything:

```
mlflow.set_experiment("PyTorch_MNIST")
```

```
with mlflow.start_run():

    preds = model.forward(torch.Tensor(x_test.float()).cuda())
    preds = np.round(preds.detach().cpu().numpy())

    eval_acc = accuracy_score(y_test, preds)
    auc_score = roc_auc_score(y_test, preds)

    mlflow.log_param("batch_size", batch_size)
    mlflow.log_param("num_epochs", num_epochs)
    mlflow.log_param("learning_rate", learning_rate)
```

```
mlflow.log_metric("eval_acc", eval_acc)
mlflow.log_metric("auc_score", auc_score)

print("eval_acc: ", eval_acc)
print("auc_score: ", auc_score)

mlflow.pytorch.log_model(model, "PyTorch_MNIST")
mlflow.end_run()
```

As you can see, MLFlow integration is still quite easy with PyTorch. Refer to Figure 4-76 to see the output.

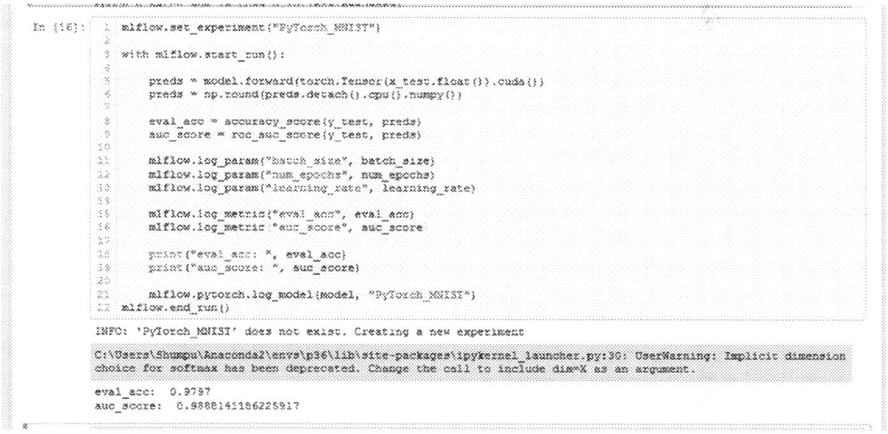

Figure 4-76. *Setting the experiment, and logging the parameters, metrics, and the model to the MLFlow run*

MLFlow UI – Checking Your Run

Let's open up the UI. Refer to Figure 4-77.

Figure 4-77. *Looking at the MLFlow UI now. Notice that your experiment, PyTorch_MNIST, is created*

As you can see, there is a new experiment titled PyTorch_MNIST. Click it. You should now see the run you just completed. Refer to Figure 4-78.

Figure 4-78. *The MLFlow UI showing your completed run*

Now that your run has shown up, click it. You should see all the parameters and metrics logged in that run. Refer to Figure 4-79.

Figure 4-79. *All the parameters, metrics, and artifacts (the model) you specified have been logged*

Also notice the model that's been saved by MLFlow under artifacts. Refer to Figure 4-80.

Figure 4-80. *MLFlow has successfully logged the model as well*

Loading an MLFlow Model

Let's now go over how to load this model using MLFlow. Copy the run ID, and head back to the notebook. Run the following, but replace the placeholders with your run ID:

```
loaded_model = mlflow.pytorch.load_model("runs:/YOUR_RUN_ID/
PyTorch_MNIST")
```

In our case, our run ID was 094a9f92cd714711926114b4c96f6d73, so our code looks like Figure 4-81.

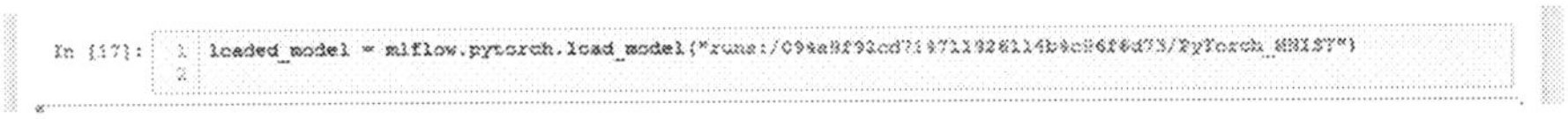

Figure 4-81. *Loading the logged MLFlow model*

Now that's done, so let's make predictions and calculate the metrics again:

```
preds = loaded_model.forward(torch.Tensor(x_test.float()).
cuda())
preds = np.round(preds.detach().cpu().numpy())
eval_acc = accuracy_score(y_test, preds)
auc_score = roc_auc_score(y_test, preds)

print("eval_acc: ", eval_acc)
print("auc_score: ", auc_score)
```

Refer to Figure 4-82 to see the output.

```
In [18]:  1  preds = loaded_model.forward(torch.Tensor(x_test.float()).cuda())
          2  preds = np.round(preds.detach().cpu().numpy())
          3  eval_acc = accuracy_score(y_test, preds)
          4  auc_score = roc_auc_score(y_test, preds)
          5
          6  print("eval_acc: ", eval_acc)
          7  print("auc_score: ", auc_score)

eval_acc:  0.9797
auc_score:  0.9888141186225917

C:\Users\Shumpu\Anaconda2\envs\p36\lib\site-packages\ipykernel_launcher.py:30: UserWarning: Implicit dimension
choice for softmax has been deprecated. Change the call to include dim=X as an argument.
```

Figure 4-82. *The output of calculating the evaluation metrics from earlier but with the logged model. As you can see, the scores match exactly*

As you can see, these metrics are the same as from the training run. You now know how to load a PyTorch model using MLFlow and how you can use it to make predictions.

With that, you now know how to integrate MLFlow into your PyTorch experiments. Next, we will look at how you can integrate MLFlow into PySpark.

MLFlow with PySpark

In our final example, we will look at how MLFlow integrates with PySpark. Like in the scikit-learn example, we will be looking at the application of a logistic regression model to the credit card dataset. In fact, this code is very similar to the PySpark example from Chapter 2.

Before we begin, here are the versions of the modules we are using, including CUDA and CuDNN:

- **PySpark** – 2.4.5
- **Matplotlib** – 3.2.1
- **Sklearn** – 0.22.2.post1
- **MLFlow** – 1.10.0
- **mumpy** – 1.18.5

Data Processing

With that, let's get started. First, you must import all the necessary modules
and set up some variables for Spark:

```python
import pyspark #
from pyspark.sql import SparkSession
from pyspark import SparkConf, SparkContext
from pyspark.sql.types import *
from pyspark.ml.feature import VectorAssembler
from pyspark.ml import Pipeline
from pyspark.ml.classification import LogisticRegression
import pyspark.sql.functions as F
import os
import seaborn as sns
import sklearn #
from sklearn.metrics import confusion_matrix
from sklearn.metrics import roc_auc_score, accuracy_score

import matplotlib #
import matplotlib.pyplot as plt

import mlflow
import mlflow.spark

os.environ["SPARK_LOCAL_IP"]='127.0.0.1'
spark = SparkSession.builder.master("local[*]").getOrCreate()
spark.sparkContext._conf.getAll()

print("pyspark: {}".format(pyspark.__version__))
print("matplotlib: {}".format(matplotlib.__version__))
print("seaborn: {}".format(sns.__version__))
print("sklearn: {}".format(sklearn.__version__))
print("mlflow: {}".format(mlflow.__version__))
```

Refer to Figure 4-83.

```
In [1]:  1  import pyspark #
         2  from pyspark.sql import SparkSession
         3  from pyspark import SparkConf, SparkContext
         4  from pyspark.sql.types import *
         5  from pyspark.ml.feature import VectorAssembler
         6  from pyspark.ml import Pipeline
         7  from pyspark.ml.classification import LogisticRegression
         8  import pyspark.sql.functions as F
         9  import os
        10  import seaborn as sns
        11  import sklearn #
        12  from sklearn.metrics import confusion_matrix
        13  from sklearn.metrics import roc_auc_score, accuracy_score
        14
        15  import matplotlib #
        16  import matplotlib.pyplot as plt
        17
        18  import mlflow
        19  import mlflow.spark
        20
        21  os.environ["SPARK_LOCAL_IP"]='127.0.0.1'
        22  spark = SparkSession.builder.master("local[*]").getOrCreate()
        23  spark.sparkContext._conf.getAll()
        24
        25  print("pyspark: {}".format(pyspark.__version__))
        26  print("matplotlib: {}".format(matplotlib.__version__))
        27  print("seaborn: {}".format(sns.__version__))
        28  print("sklearn: {}".format(sklearn.__version__))
        29  print("mlflow: {}".format(mlflow.__version__))

         pyspark: 2.4.5
         matplotlib: 3.2.1
         seaborn: 0.10.1
         sklearn: 0.22.2.post1
         mlflow: 1.10.0
```

Figure 4-83. *Importing the necessary modules and printing their versions*

Next, let's load your data set and specify what columns you want to take:

```
data_path = 'data/creditcard.csv'

df = spark.read.csv(data_path, header = True,
inferSchema = True)
labelColumn = "Class"
columns = df.columns
numericCols = columns
numericCols.remove("Time")
numericCols.remove(labelColumn)
print(numericCols)
```

Refer to Figure 4-84 to see the output.

```
In [2]:    1  data_path = 'data/creditcard.csv'
           2
           3  df = spark.read.csv data_path, header = True, inferSchema = True
           4  labelColumn = "Class"
           5  columns = df.columns
           6  numericCols = columns
           7  numericCols.remove("Time")
           8  numericCols.remove(labelColumn)
           9  print(numericCols)

           ['V1', 'V2', 'V3', 'V4', 'V5', 'V6', 'V7', 'V8', 'V9', 'V10', 'V11', 'V12', 'V13', 'V14', 'V15', 'V16', 'V17',
           'V18', 'V19', 'V20', 'V21', 'V22', 'V23', 'V24', 'V25', 'V26', 'V27', 'V28', 'Amount']
```

Figure 4-84. *Loading the data and specifying the columns that you want as a list*

Notice that you dropped the column Time here, like with the scikit-learn example. This column just adds a lot of extraneous information that doesn't actually correlate very much with the label column and could even possibly make the learning task harder than it needs to be.

Let's see what the data frame looks like:

```
df.toPandas().head()
```

Refer to Figure 4-85 to see the output.

```
In [3]:    1  df.toPandas().head()

Out[3]:
        Time      V1        V2        V3        V4        V5        V6        V7        V8        V9   ...      V21       V22       V23       V24      \
   0      0  -1.359807  -0.072781  2.536347  1.378155  -0.338321  2.462388  0.239599  0.098698  0.363787  ...  -0.018307  0.277838  -0.110474  0.066928  8.128(
   1      0   1.191857  0.266151  0.166480  0.448154  0.060018  -0.082361  -0.078803  0.085102  -0.255425  ...  -0.225775  -0.638672  0.101288  -0.339846  0.187
   2      1  -1.358354  -1.340163  1.773209  0.379780  -0.503198  1.800499  0.791461  0.247676  -1.514654  ...  0.247998  0.771679  0.909412  -0.689281  -8.327
   3      1  -0.966272  -0.185226  1.792993  -0.863291  -0.010309  1.247203  0.237609  0.377436  -1.387024  ...  -0.108300  0.005274  -0.190321  -1.175575  0.647
   4      2  -1.158233  0.877737  1.548718  0.403034  -0.407193  0.095921  0.592941  -0.270533  0.817739  ...  -0.009431  0.798278  -0.137458  0.141267  -0.206(

5 rows × 31 columns
```

Figure 4-85. *Converting the Spark data frame to Pandas and checking the output. As you can see, the columns have loaded in correctly, along with the data. The column Time has not been dropped because you did not filter the data frame yet*

You'll notice that the columns you "dropped" are still showing up, like Time. You haven't filtered the columns you want yet, which you are going to do now. Run the following to select the features you want from the data frame and create your normal and anomaly splits:

```
stages = []
assemblerInputs =  numericCols
assembler = VectorAssembler(inputCols=assemblerInputs,
outputCol="features")
stages += [assembler]

dfFeatures = df.select(F.col(labelColumn).alias('label'),
*numericCols )

normal = dfFeatures.filter("Class == 0").
sample(withReplacement=False, fraction=0.5, seed=2020)
anomaly = dfFeatures.filter("Class == 1")

normal_train, normal_test = normal.randomSplit([0.8, 0.2],
seed = 2020)
anomaly_train, anomaly_test = anomaly.randomSplit([0.8, 0.2],
seed = 2020)
```

Refer to Figure 4-86 to see the code in a cell.

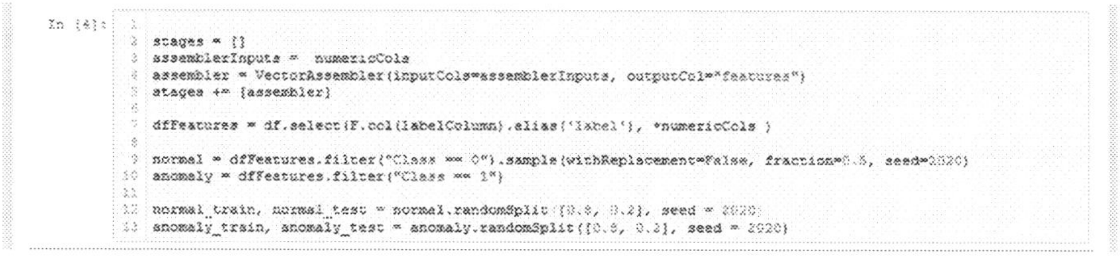

Figure 4-86. *Selecting the columns that you want and defining your normal and anomaly train and test sets*

Let's look at the new data frame now:

```
dfFeatures.toPandas().head()
```

Refer to Figure 4-87.

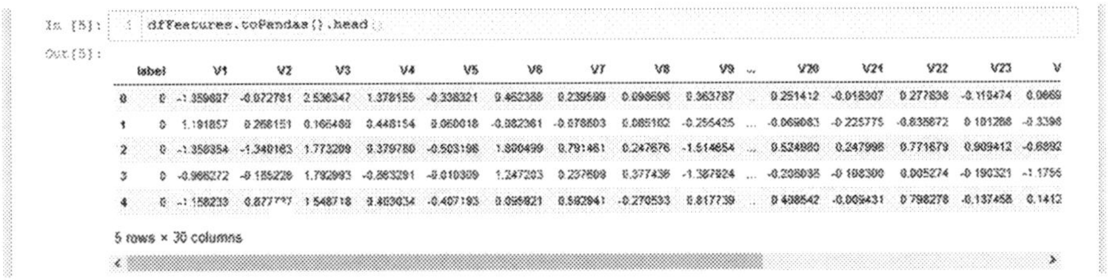

	label	V1	V2	V3	V4	V5	V6	V7	V8	V9	...	V20	V21	V22	V23	V
0	0	-1.359807	-0.072781	2.536347	1.378155	-0.338321	0.462388	0.239599	0.098698	0.363787	...	0.251412	-0.018307	0.277838	-0.110474	0.0669
1	0	1.191857	0.266151	0.166480	0.448154	0.060018	-0.082361	-0.078803	0.085102	-0.255425	...	-0.069083	-0.225775	-0.638672	0.101288	-0.3398
2	0	-1.358354	-1.340163	1.773209	0.379780	-0.503198	1.800499	0.791461	0.247676	-1.514654	...	0.524980	0.247998	0.771679	0.909412	-0.6893
3	0	-0.966272	-0.185226	1.792993	-0.863291	-0.010309	1.247203	0.237609	0.377436	-1.387024	...	-0.208038	-0.108300	0.005274	-0.190321	-1.1756
4	0	-1.158233	0.877737	1.548718	0.403034	-0.407193	0.095921	0.592941	-0.270533	0.817739	...	0.408542	-0.009431	0.798278	-0.137458	0.1412

5 rows × 30 columns

Figure 4-87. *As you can see, Time has been dropped. This is the data frame that your training and testing sets are derived from*

Notice that the columns you dropped are gone. Now you know that normal and anomaly don't have the features you dropped either and that everything is proceeding as planned. Let's create the train and test sets:

```
train_set = normal_train.union(anomaly_train)
test_set = normal_test.union(anomaly_test)
```

Refer to Figure 4-88.

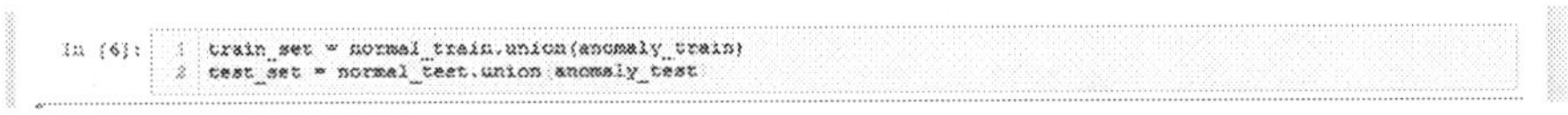

Figure 4-88. *Concatenating the normal and anomaly sets to create the train and test sets*

Let's now move on to creating the feature vector that the logistic regression model is going to use. Run the following to define the pipeline and create your final train and test sets:

```
pipeline = Pipeline(stages = stages)
pipelineModel = pipeline.fit(dfFeatures)
train_set = pipelineModel.transform(train_set)
```

```
test_set = pipelineModel.transform(test_set)
selectedCols = ['label', 'features'] + numericCols
train_set = train_set.select(selectedCols)
test_set = test_set.select(selectedCols)

print("Training Dataset Count: ", train_set.count())
print("Test Dataset Count: ", test_set.count())
```

Refer to Figure 4-89.

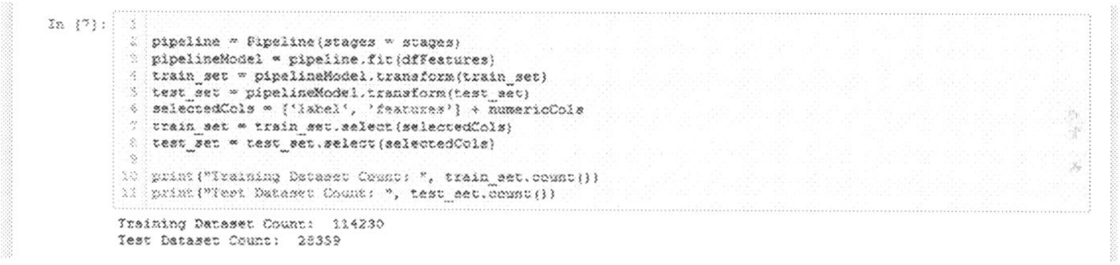

```
In [7]:  1
         2   pipeline = Pipeline(stages = stages)
         3   pipelineModel = pipeline.fit(dfFeatures)
         4   train_set = pipelineModel.transform(train_set)
         5   test_set = pipelineModel.transform(test_set)
         6   selectedCols = ['label', 'features'] + numericCols
         7   train_set = train_set.select(selectedCols)
         8   test_set = test_set.select(selectedCols)
         9
        10   print("Training Dataset Count: ", train_set.count())
        11   print("Test Dataset Count: ", test_set.count())

Training Dataset Count:  114230
Test Dataset Count:  28359
```

Figure 4-89. *Defining the pipeline used to create the feature vector that will be used to train the model. From the feature vector and the label vector, you define your final train and test sets*

Now that you've finished processing the data, let's define a function to train the model and calculate some relevant metrics:

```
def train(spark_model, train_set):

    trained_model = spark_model.fit(train_set)

    trainingSummary = trained_model.summary
    pyspark_auc_score = trainingSummary.areaUnderROC

    mlflow.log_metric("train_acc", trainingSummary.accuracy)
    mlflow.log_metric("train_AUC", pyspark_auc_score)

    print("Training Accuracy: ", trainingSummary.accuracy)
    print("Training AUC:", pyspark_auc_score)

    return trained_model
```

Refer to Figure 4-90 to see the function in a cell.

```
In [8]:   1  def train(spark_model, train_set):
          2
          3      trained_model = spark_model.fit(train_set)
          4
          5      trainingSummary = trained_model.summary
          6      pyspark_auc_score = trainingSummary.areaUnderROC
          7
          8      mlflow.log_metric("train_acc", trainingSummary.accuracy)
          9      mlflow.log_metric("train_AUC", pyspark_auc_score)
         10
         11      print("Training Accuracy: ", trainingSummary.accuracy)
         12      print("Training AUC:", pyspark_auc_score)
         13
         14      return trained_model
```

Figure 4-90. *The code to train the PySpark logistic regression model and log the training accuracy and AUC score metrics*

Let's now define a function to evaluate the model and calculate those metrics, too:

```
def evaluate(spark_model, test_set):

    evaluation_summary = spark_model.evaluate(test_set)

    eval_acc = evaluation_summary.accuracy
    eval_AUC = evaluation_summary.areaUnderROC

    mlflow.log_metric("eval_acc", eval_acc)
    mlflow.log_metric("eval_AUC", eval_AUC)

    print("Evaluation Accuracy: ", eval_acc)
    print("Evaluation AUC: ", eval_AUC)
```

Refer to Figure 4-91.

```
In [9]:   1  def evaluate(spark_model, test_set):
          2
          3      evaluation_summary = spark_model.evaluate(test_set)
          4
          5      eval_acc = evaluation_summary.accuracy
          6      eval_AUC = evaluation_summary.areaUnderROC
          7
          8      mlflow.log_metric("eval_acc", eval_acc)
          9      mlflow.log_metric("eval_AUC", eval_AUC)
         10
         11      print("Evaluation Accuracy: ", eval_acc)
         12      print("Evaluation AUC: ", eval_AUC)
         13
```

Figure 4-91. *The code to evaluate the trained PySpark logistic regression model and log the evaluation accuracy and AUC score metrics*

MLFlow Run – Training, UI, and Loading an MLFlow Model

Now that you have finished defining the training and evaluation functions along with the metrics you want to log, it's time to start an MLFlow run and build a model:

```
lr = LogisticRegression(featuresCol = 'features', labelCol = 'label', maxIter=10)

mlflow.set_experiment("PySpark_CreditCard")

with mlflow.start_run():

    trainedLR = train(lr, train_set)

    evaluate(trainedLR, test_set)

    mlflow.spark.log_model(trainedLR, "creditcard_model_pyspark")

mlflow.end_run()
```

207

Refer to Figure 4-92.

```
In [10]:    1  lr = LogisticRegression(featuresCol = 'features', labelCol = 'label', maxIter=10)
            2
            3  mlflow.set_experiment("PySpark_CreditCard")
            4
            5  with mlflow.start_run():
            6
            7
            8      trainedLR = train(lr, train_set)
            9
           10      evaluate(trainedLR, test_set)
           11
           12      mlflow.spark.log_model(trainedLR, "creditcard_model_pyspark")
           13
           14  mlflow.end_run()

INFO: 'PySpark_CreditCard' does not exist. Creating a new experiment
Training Accuracy:   0.9986680714348244
Training AUC:  0.9778117954354749
Evaluation Accuracy:   0.998624775203639
Evaluation AUC:   0.9788891095270811
```

Figure 4-92. *The output of the MLFlow run. The experiment has been created and the metrics and model successfully logged*

Alright, now that MLFlow has finished logging everything and the run has ended, open up the MLFlow UI. You should see something like Figure 4-93.

Figure 4-93. *The MLFlow UI showing that your experiment, PySpark_CreditCard, has been created*

Notice that a new experiment called PySpark_CreditCard has been created. Click it, and you should see something like Figure 4-94. If MLFlow didn't log the run here, try rerunning the cell. It should log it correctly.

Figure 4-94. *MLFlow UI showing that your run has successfully finished*

If everything went well, you should see a run logged in this experiment. Click it, and you should see something like Figure 4-95.

Figure 4-95. *Looking at the run, it appears that all of your metrics have successfully been logged*

Finally, in the artifacts section, click the folder that says creditcard_model_pyspark to expand it. You should see a folder called sparkml that contains the PySpark logistic regression model. Refer to Figure 4-96.

Figure 4-96. *MLFlow has also logged the PySpark model. There is no concrete model file like with the TensorFlow or PyTorch examples because of the way PySpark stores its models*

Now that you've verified MLFlow has logged everything you specified, copy the run number at the top. Now go back to the notebook and run the following, replacing the placeholder with your run:

```
model = mlflow.spark.load_model("runs:/YOUR_RUN_ID/
creditcard_model_pyspark")
```

In our case, our run was 58e6aac5d43948c6948bee29c0c04cca, so our cell looks like Figure 4-97.

```
In [11]:   1 model = mlflow.spark.load_model("runs:/58e6aac5d43948c6948bee29c0c04cca/creditcard_model_pyspark")

2020/08/02 18:15:59 INFO mlflow.spark: 'runs:/58e6aac5d43948c6948bee29c0c04cca/creditcard_model_pyspark' resol
ved as 'file:///C:/Users/Shumpu/work/Books/2020%20MLOps/Chapter%204/mlruns/14/58e6aac5d43948c6948bee29c0c04cca
/artifacts/creditcard_model_pyspark'
2020/08/02 18:15:59 INFO mlflow.spark: File 'file:///C:/Users/Shumpu/work/Books/2020%20MLOps/Chapter%204/mlrun
s/14/58e6aac5d43948c6948bee29c0c04cca/artifacts/creditcard_model_pyspark/sparkml' not found on DFS. Will attem
pt to upload the file.
2020/08/02 18:16:01 INFO mlflow.spark: Copied SparkML model to /tmp/mlflow\102e1969-96bc-4144-90b3-6fa66246897
1
```

Figure 4-97. *Loading the logged MLFlow model*

Now that the model has been loaded, let's make some predictions with it. Run the following:

```
predictions = model.transform(test_set)
y_true = predictions.select(['label']).collect()
y_pred = predictions.select(['prediction']).collect()
```

Refer to Figure 4-98 to see the code in a cell.

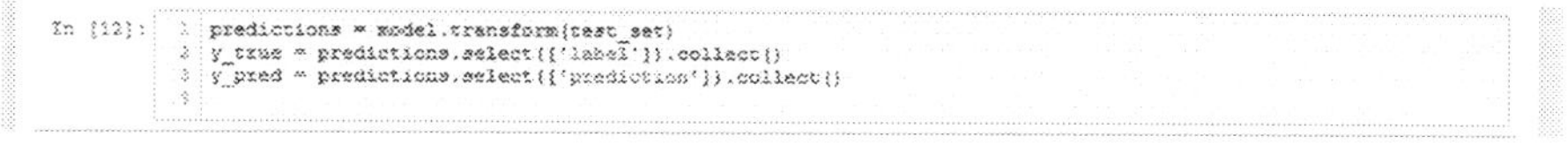

```
In [12]:    1  predictions = model.transform(test_set)
            2  y_true = predictions.select(['label']).collect()
            3  y_pred = predictions.select(['prediction']).collect()
            4
```

Figure 4-98. *Making predictions with your loaded model*

Let's print out the evaluation accuracy and the AUC score:

```
print(f"AUC Score: {roc_auc_score(y_true, y_pred):.3%}")
print(f"Accuracy Score: {accuracy_score(y_true, y_pred):.3%}")
```

Refer to Figure 4-99.

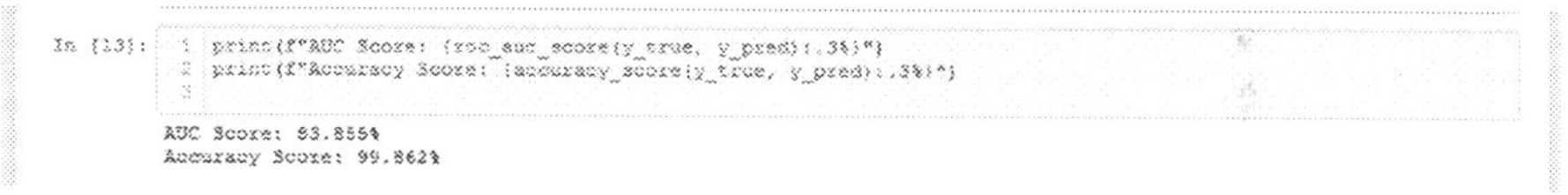

```
In [13]:    1  print(f"AUC Score: {roc_auc_score(y_true, y_pred):.3%}")
            2  print(f"Accuracy Score: {accuracy_score(y_true, y_pred):.3%}")
            3

AUC Score: 83.855%
Accuracy Score: 99.862%
```

Figure 4-99. *Printing out the evaluation metrics. The AUC score noticeably differs, but the accuracy score matches what was displayed during the MLFlow run*

You will notice that the AUC score differs compared to what was calculated in the evaluation function. This is likely because PySpark calculates the ROC curve slightly differently because it has direct access to the model itself. On the other hand, with scikit-learn, you only have the true labels and the predictions to work with, so the ROC curve is calculated slightly differently.

Finally, let's construct the confusion matrix:

```
conf_matrix = confusion_matrix(y_true, y_pred)
ax = sns.heatmap(conf_matrix, annot=True, fmt='g')
ax.invert_xaxis()
ax.invert_yaxis()
plt.ylabel('Actual')
plt.xlabel('Predicted')
```

Refer to Figure 4-100.

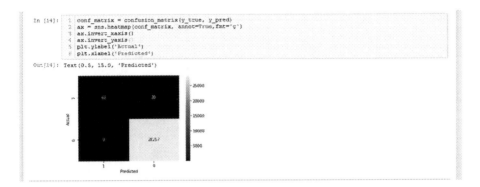

Figure 4-100. *Displaying the confusion matrix using the true values and the predictions made by the model you loaded*

From the confusion matrix, you can see that the AUC score as calculated by PySpark must be reflecting its performance on how well it classifies normal data. Looking at the anomalies, a fair chunk of the fraudulent data has been misclassified. Roughly speaking, the model only got two-thirds of the anomalies when evaluated on the test data. Perhaps this explains the disparity between what scikit-Learn says is the AUC score and what PySpark says is the AUC score. Both must have calculated the ROC curves slightly differently with PySpark's graph somehow favoring the excellent true positive rate of the normal data's classification.

With that, you now know how to integrate MLFlow into your PySpark experiments.

Next, we will take a look at how you can deploy your models locally and how you can query the models with samples of data and receive predictions.

Local Model Serving

Deploying the Model

Serving and querying models locally is very easy and can be done in the command line. You only need the experiment ID and the run ID to serve the model. This is where the print statement from earlier can apply, as it prints the run ID of that specific run. If you just want to serve the latest model, you may do so using that ID.

Otherwise, you can look in the MLFlow UI, select a model run that suits your needs, and paste the run this way.

Before you begin, go to the MLFlow UI once again, and click the experiment scikit_learn_experiment. Pick a run and copy the run ID. Don't forget the model name that you logged the model with either, which should be log_reg_model.

You may create a new notebook at this point to keep the code more organized, but be sure to import the following:

```
import pandas as pd
import mlflow
import mlflow.sklearn

import seaborn as sns

import matplotlib.pyplot as plt

from sklearn.preprocessing import StandardScaler
from sklearn.model_selection import train_test_split
from sklearn.metrics import roc_auc_score, accuracy_score,
confusion_matrix
```

```
import numpy as np

import subprocess
import json
```

You'll notice that you are now importing subprocess. If you're using the same notebook, make sure to import this module as well.

Refer to Figure 4-101 to see this code in a cell.

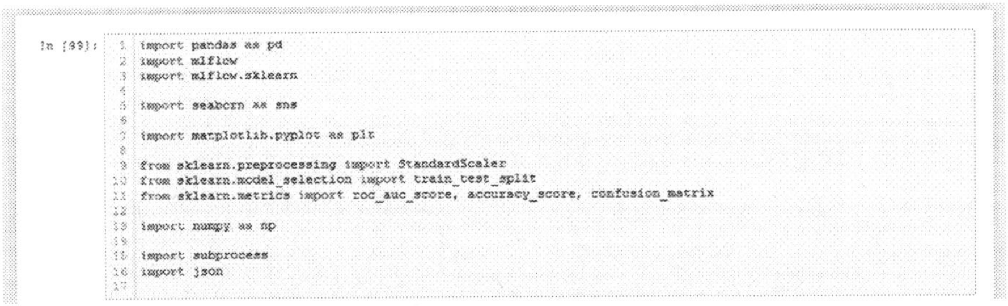

Figure 4-101. *Importing the necessary modules*

Now, open up your command prompt/terminal so that you can begin to serve your local model. First, you need to change your directory to one that contains the `mlruns` folder with all your experiments. Next, you need two things: your **model run** and your **model name**.

Again, your model run can be anything you pick from the MLFlow UI or it can simply be the latest run. The model name is whatever you set it to when logging the model. In this case, it will be `log_reg_model`.

Once you have that, run the following command in your command prompt/terminal. We have generalized the command, so be sure to replace the fields with your model run and model name, respectively:

```
mlflow models serve --model-uri runs:/YOUR_MODEL_RUN/
YOUR_MODEL_NAME -p 1235
```

In our case, our model run was `3862eb3bd89b43e8ace610c521d974e6`, and our model name was once again `log_reg_model`. And so, the command we ran looks like Figure 4-102.

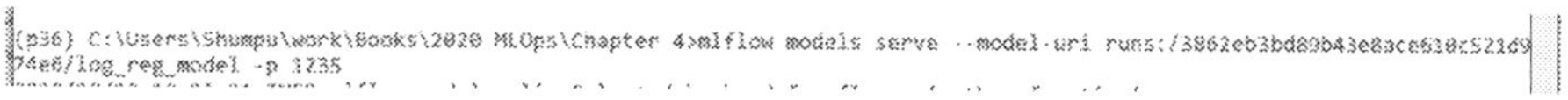

```
(p36) C:\Users\Shumpu\work\Books\2020 MLOps\Chapter 4>mlflow models serve --model-uri runs:/3862eb3bd89b43e8ace610c521d9
74e6/log_reg_model -p 1235
```

Figure 4-102. *The command that we ran to serve our model locally*

In text, the command looks like this:

```
mlflow models serve --model-uri runs:/3862eb3bd89b43e8ace610c52
1d974e6/log_reg_model -p 1235
```

MLFlow should start constructing a new conda environment right away that it will use to serve locally. In this environment, it installs basic packages and specific packages that the model needs to be able to run.

After some time, you should see something like in Figure 4-103.

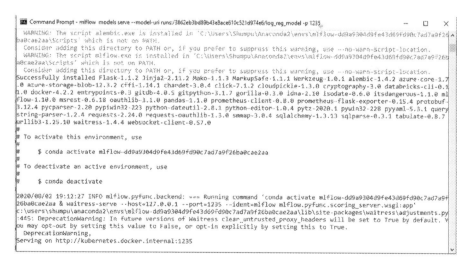

Figure 4-103. *The result of running the command to deploy the model locally. You might see something different, such as localhost:1235, but this is because we have docker installed*

MLFlow should create a new conda environment before hosting the model on your local server. The port option -p lets you set a specific port to host the model on. We selected a specific port so that we can have MLFlow UI running at the same time, as both of them default to port 5000. In our case, our MLFlow UI is running on port 1234, so we are serving the model on port 1235.

Querying the Model

You are now ready to query the model with data and receive predictions. This is where the subprocess module comes in, and you'll see why shortly. First, let's load up your data frame again. Run the following code:

```
df = pd.read_csv("data/creditcard.csv")
```

You should see something like Figure 4-104.

```
In [3]:   1   df = pd.read_csv("data/creditcard.csv"
```

Figure 4-104. *Loading the credit card dataset*

Next, select 80 values from your data frame to query your model with. Run the following code:

```
input_json = df.iloc[:80].drop(["Time", "Class"],
axis=1).to_json(orient="split")
```

You should see something like Figure 4-105.

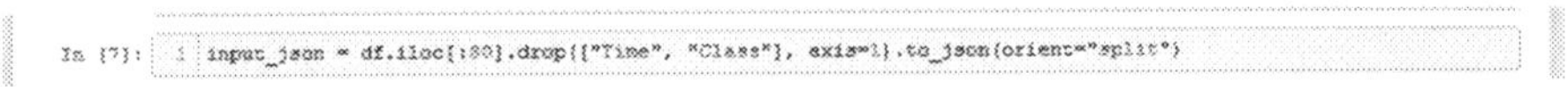

```
In [7]:   1   input_json = df.iloc[:80].drop(["Time", "Class"], axis=1).to_json(orient="split")
```

Figure 4-105. *Converting a selection of 80 rows, dropping the Time and Class columns since they were dropped in the original x_train used to train the model, to a JSON with a split orient*

216

The next step is important because of how you preprocessed the data before training your model originally. To show why it's so important, we will quickly demonstrate the difference in evaluation metrics from passing in non-scaled data and scaled data. First of all, here is the code to send data to the model and receive predictions back:

```
proc = subprocess.run(["curl",  "-X", "POST", "-H",
"Content-Type:application/json; format=pandas-split",
"--data", input_json, "http://127.0.0.1:1235/invocations"],
stdout=subprocess.PIPE, encoding='utf-8')

output = proc.stdout
df2 = pd.DataFrame([json.loads(output)])
df2
```

Essentially, what this does is run the following command within Python itself:

```
curl -X POST -H "Content-Type:application/json;
format=pandas-split" -data "CONTENT_OF_INPUT_JSON"
"http://127.0.0.1:1235/invocations"
```

The core of the problem is that if you are running this in command line, pasting the JSON format data of the data frame can get very messy because there's so many columns. That is why we chose to use subprocess as it is easier to directly pass in the JSON itself using a variable name, input_json in this case, to hold the contents of the JSON.

You should see something like Figure 4-106.

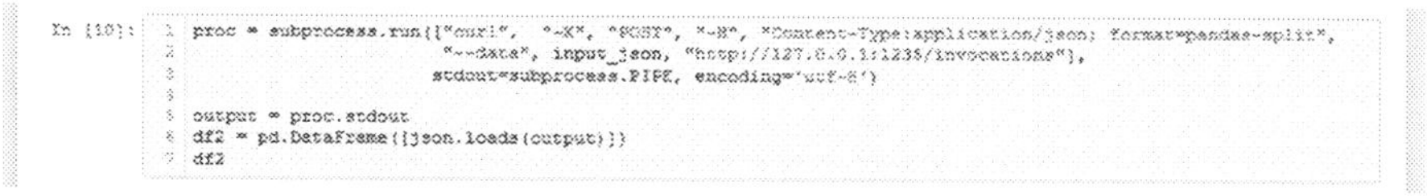

Figure 4-106. *Sending data to the locally hosted model and receiving predictions from the model*

Now, you will query the model with input data that is not scaled.

Querying Without Scaling

You will keep the selection of 80 values from earlier and query the model. The model accepts data in the JSON format, so you will have to convert the format of your data before sending it to the model. Run the cell in Figure 4-106.

You should see something like Figure 4-107.

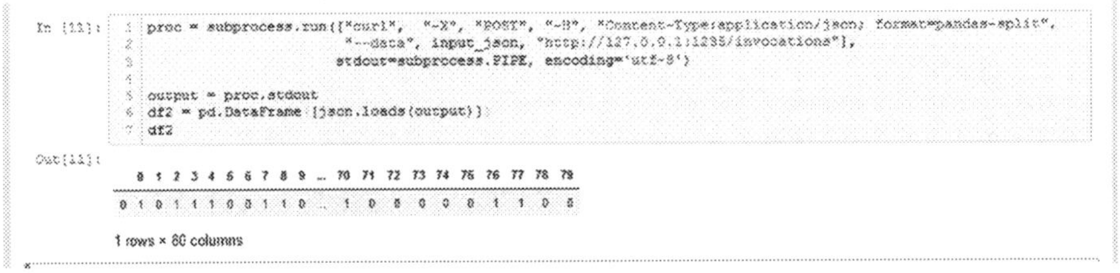

Figure 4-107. *The list of predictions that you get after querying the model with input_json. Notice that it's predicting a lot of anomalies. This is the first red flag that indicates something's wrong*

The resulting data frame is what you get by converting the predictions that you got back from the model into a data frame. Since you have the true predictions, let's calculate an AUC score and an accuracy score to see how the model did. Run the following code:

```
y_true = df.iloc[:80].Class
df2 = df2.T
eval_acc = accuracy_score(y_true, df2)

y_true.iloc[-1] = 1
eval_auc = roc_auc_score(y_true, df2)

print("Eval Acc", eval_acc)
print("Eval AUC", eval_auc)
```

First of all, you had to transpose df2 using .T so that you can get the predictions to be in a Pandas Series format. Next, the AUC score cannot be calculated if one of the arrays y_true or y_preds only have one class. In this case, y_true is only comprised of normal values, so you had to manipulate the last value and make it 1 when it really isn't just to get an AUC score. Of course, the resulting AUC score will be nonsense.

You should see something like Figure 4-108.

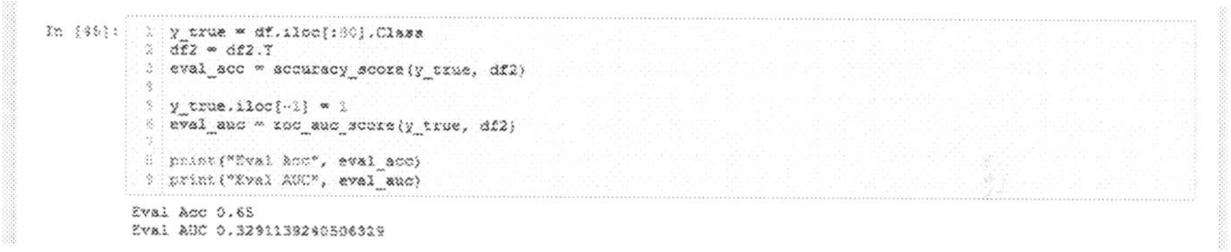

```
In [95]:  1  y_true = df.iloc[:80].Class
          2  df2 = df2.T
          3  eval_acc = accuracy_score(y_true, df2)
          4
          5  y_true.iloc[-1] = 1
          6  eval_auc = roc_auc_score(y_true, df2)
          7
          8  print("Eval Acc", eval_acc)
          9  print("Eval AUC", eval_auc)

Eval Acc 0.65
Eval AUC 0.3291139240506329
```

Figure 4-108. *Evaluating the accuracy and the AUC score from the predictions. The AUC score is nonsense, but the accuracy score reveals that the model has performed very poorly*

As you can see, the accuracy score is horrible. This basically means that the model doesn't know the difference between the anomalies and the normal points but seems to have some idea about normal points.

The reason the model did so poorly despite doing so well during the training process is that the input data has not been scaled. You will see the difference in model performance when you now scale the data before passing it in.

219

Querying with Scaling

You will take the same split of data except you will now scale it before passing it in. Run the following code to recreate the data that you used to fit the scaler when training the model originally:

```
normal = df[df.Class == 0].sample(frac=0.5, random_state=2020).
reset_index(drop=True)
anomaly = df[df.Class == 1]

normal_train, normal_test = train_test_split(normal,
test_size = 0.2, random_state = 2020)
anomaly_train, anomaly_test = train_test_split
(anomaly, test_size = 0.2, random_state = 2020)

scaler = StandardScaler()
scaler.fit(pd.concat((normal, anomaly)).drop(["Time", "Class"],
axis=1))
```

You should see something like Figure 4-109.

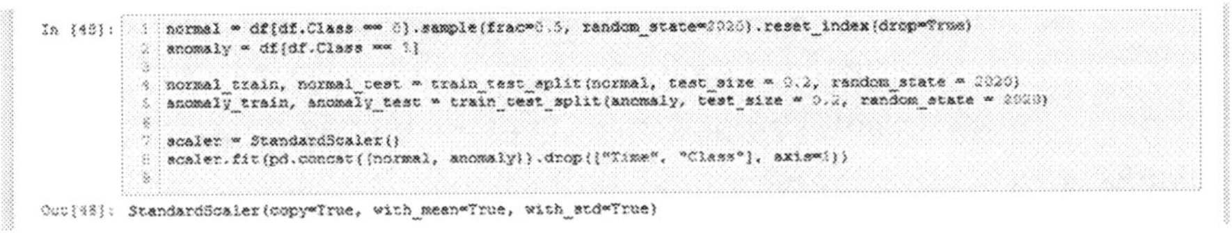

Figure 4-109. *Recreating the original dataset that you used to fit the standard scaler when processing the data originally. Using this, you will transform your new sample of data and pass it into the model*

Now that you have fit the scaler, let's transform your data selection:

```
scaled_selection = scaler.transform(df.iloc[:80].drop
(["Time", "Class"], axis=1))
input_json = pd.DataFrame
(scaled_selection).to_json(orient="split")
```

Refer to Figure 4-110.

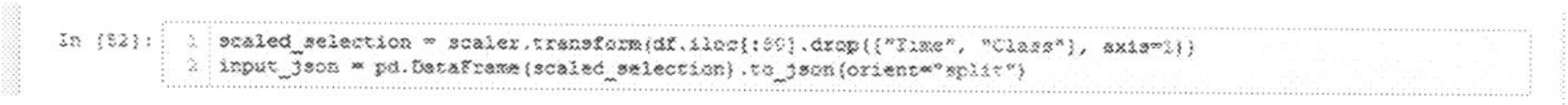

```
In [52]:   1  scaled_selection = scaler.transform(df.iloc[:80].drop(["Time", "Class"], axis=1))
           2  input_json = pd.DataFrame(scaled_selection).to_json(orient="split")
```

Figure 4-110. *Scaling the selection of 80 values from the original data frame and converting it into a JSON format to be sent to the model*

Now run the following:

```
proc = subprocess.run(["curl",   "-X", "POST", "-H",
        "Content-Type:application/json; format=pandas-split",
        "--data", input_json, "http://127.0.0.1:1235/invocations"],
        stdout=subprocess.PIPE, encoding='utf-8')

output = proc.stdout
preds = pd.DataFrame([json.loads(output)])
preds
```

You should see something like Figure 4-111.

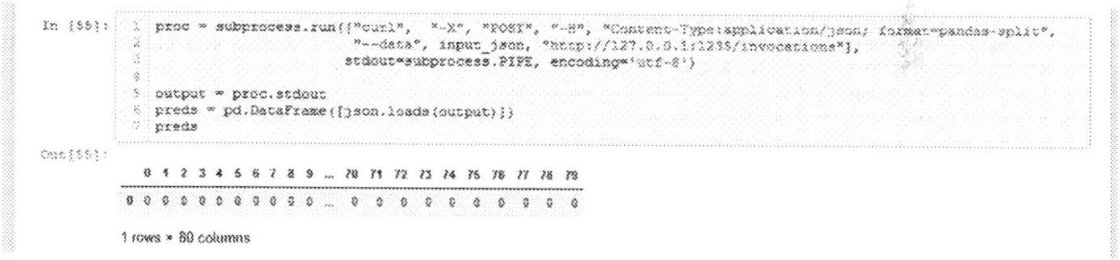

```
In [55]:   1  proc = subprocess.run(["curl",   "-X", "POST", "-H", "Content-Type:application/json; format=pandas-split",
                    "--data", input_json, "http://127.0.0.1:1235/invocations"],
                    stdout=subprocess.PIPE, encoding='utf-8')
           5  output = proc.stdout
           6  preds = pd.DataFrame([json.loads(output)])
           7  preds
Out[55]:
         0  1  2  3  4  5  6  7  8  9  ... 70 71 72 73 74 75 76 77 78 79
      0  0  0  0  0  0  0  0  0  0  0  ...  0  0  0  0  0  0  0  0  0  0

1 rows × 80 columns
```

Figure 4-111. *Querying the model with the scaled values. From a first glance, the predictions appear to be correct this time around*

One thing to note is that you are scaling it on the combination of all normal data and all anomaly data, as you did when you were creating the train, test, and validation splits. Since the model was trained on data that was scaled on the partition of data you used in the training process (the training, testing, and validation data together), passing in data scaled differently won't result in the correct predictions. When you scale the new data, it must be scaled after fitting it on the training set.

One problem that may eventually arise is that new data might have a different distribution than the original training data. This could lead to performance issues with the model, but really that's a sign that you need to train your model to update it on the new data.

Let's check how your model did now:

```
y_true = df.iloc[:80].Class
preds = preds.T
eval_acc = accuracy_score(y_true, preds)

y_true.iloc[-1] = 1
eval_auc = roc_auc_score(y_true, preds)

print("Eval Acc", eval_acc)
print("Eval AUC", eval_auc)
```

Refer to Figure 4-112.

```
In [68]:    1 y_true = df.iloc[:80].Class
            2 preds = preds.T
            3 eval_acc = accuracy_score(y_true, preds)
            4
            5 y_true.iloc[-1] = 1
            6 eval_auc = roc_auc_score(y_true, preds)
            7
            8 print("Eval Acc", eval_acc)
            9 print("Eval AUC", eval_auc)

Eval Acc 0.9875
Eval AUC 0.5
```

Figure 4-112. *Checking the accuracy and the AUC scores of the predictions. The accuracy score is far better, but you will need more prediction data with both normal and anomaly values to be able to get AUC scores*

As you can see, the accuracy score is noticeably higher, and the model's performance is reminiscent of when it was trained and evaluated. Unfortunately, the AUC score isn't a very accurate reflection of the model's performance since the samples you are querying the model with only have normal data.

Let's see how the model performs when you query it with a larger sample of data.

Batch Querying

Unfortunately, there is a limit to how many data samples you can ask the model to make predictions on. The number 80 is really close to the maximum number of samples you can send at one time. So how do you get around this issue and make predictions on more than just 80 samples? For one, you can try batching the samples and making predictions one batch at a time.

Run the following code:

```
test = df.iloc[:8000]
true = test.Class
test = scaler.transform(test.drop(["Time", "Class"], axis=1))
preds = []

batch_size = 80
for f in range(100):
    sample = pd.DataFrame(test[f*batch_size:(f+1)*batch_size]).
    to_json(orient="split")
    proc = subprocess.run(["curl",  "-X", "POST", "-H",
                            "Content-Type:application/json;
                            format=pandas-split", "--data",
                            sample, "http://127.0.0.1:1235/
                            invocations"],
                            stdout=subprocess.PIPE,
                            encoding='utf-8')
```

```
output = proc.stdout
resp = pd.DataFrame([json.loads(output)])
preds = np.concatenate((preds, resp.values[0]))
```

```
eval_acc = accuracy_score(true, preds)
eval_auc = roc_auc_score(true, preds)
```

```
print("Eval Acc", eval_acc)
print("Eval AUC", eval_auc)
```

Here, you are selecting the first 8,000 samples from the data frame. Since the batch size is 80, you have 100 batches that you are passing to the model. Of course, you must scale this data as well before passing it in. You will scale it in a manner similar to how you did it earlier: you will fit the scaler on the same normal and anomaly data that you used in the model training pipeline samples to transform the values you want to send to the model. Once finished, you should see something like Figure 4-113. This might take several seconds to finish, so sit tight!

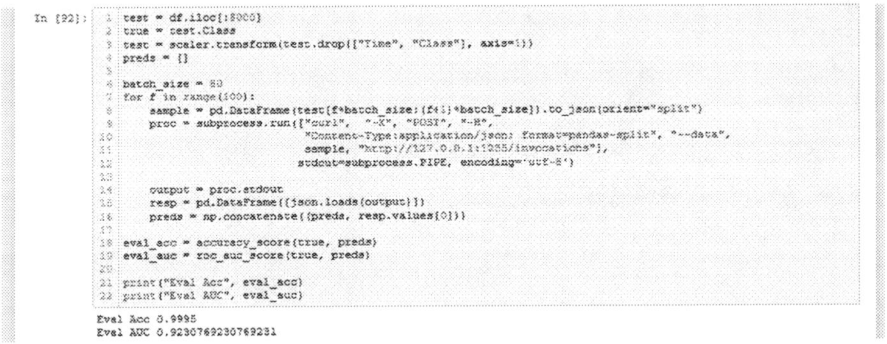

Figure 4-113. *The results of querying the model with the first 8,000 samples in the data frame. Notice that the AUC score is far better samples*

224

This time, you don't have to worry about only having one class in the entire data. This is because there are examples of anomalies in this selection of 8,000 data points, so the true labels and predictions should contain samples of both classes.

You can see that the model performs quite well on this data, which includes data that the model has never seen before. Although you did end up using all of the anomalies when training the data, the model still performs well on the normal data, as evidenced by the relatively high AUC score.

In fact, let's plot a confusion matrix to see how the model did and what's bringing down the AUC score. Run the following code:

```
conf_matrix = confusion_matrix(true, preds)
ax = sns.heatmap(conf_matrix, annot=True,fmt='g')
ax.invert_xaxis()
ax.invert_yaxis()
plt.ylabel('Actual')
plt.xlabel('Predicted')
plt.title("Confusion Matrix")
```

Refer to Figure 4-114 to see the output.

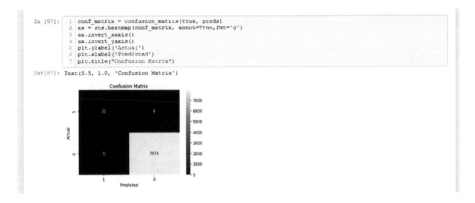

Figure 4-114. The confusion matrix for the predictions and true values. The model performed excellently and was able to classify every normal point correctly and a majority of the anomaly points correctly samples

As you can see, the confusion matrix shows that the model has performed very well on this data. Not only did it classify the normal points perfectly, but it even classified most of the anomaly points correctly as well.

With that, you hopefully know more about the process of deploying and querying a model. When you deploy to a cloud platform, the querying process follows a similar path where you must deploy a model on the cloud platform and query it by sending in the data in a JSON format.

Summary

MLFlow is an API that can help you integrate MLOps principles into your existing code base, supporting a wide variety of popular frameworks. In this chapter, we covered how you can use MLFlow to log metrics, parameters, graphs, and the models themselves. Additionally, you learned how to load the logged model and make use of its functionality. As for frameworks, we covered how you can apply MLFlow to your experiments in scikit-learn, TensorFlow 2.0/Keras, PyTorch, and PySpark, and we also looked at how you can take one of these models, deploy it locally, and make predictions with your model.

In the next chapter, we will look at how you can take your MLFlow models and use MLFlow functionality to help deploy them to Amazon SageMaker. Furthermore, we will also look at how you can make predictions using your deployed model.

CHAPTER 5

Deploying in AWS

In this chapter, we will cover how you can operationalize your MLFlow
models using AWS SageMaker. We will cover how you can upload your
runs to S3 storage, how you can build and push an MLFlow Docker
container image to AWS, and how you can deploy your model, query it,
update the model once it is deployed, and remove a deployed model.

Introduction

In the previous chapter, you learned what MLFlow is and how you can
utilize the functionality it provides to integrate MLOps principles into
your code. You also looked at how to deploy a model to a local server and
perform model inference. However, now it's time to move to the next stage
and explore how you can deploy your machine learning models to a cloud
platform so that multiple entities can use its prediction services.

Before you begin, here are some **important prerequisites**:

- You must have the AWS command line interface (CLI)
  installed and have your credentials configured.

  - Once your credentials are verified, the AWS CLI lets
    you connect to your AWS workspace. From here,
    you can create new buckets, check your SageMaker
    endpoints, and so on all through the command
    line.

© Sridhar Alla, Suman Kalyan Adari 2021
S. Alla and S. K. Adari, *Beginning MLOps with MLFlow*,
https://doi.org/10.1007/978-1-4842-6549-9_5

- You must have an Identity and Access Management (IAM) execution role defined that grants SageMaker access to your S3 buckets. Refer to Figure 5-8 to see more on this.

- You must have Docker installed and working properly. Verify that you can build Docker images.

  – It is essential to have Docker working on your system because without it, MLFlow cannot build the Docker container image to push to the AWS ECR.

We also recommend that you learn about AWS in general and how it works. Having background knowledge of AWS and how it works can help you understand this chapter and allow you to fix any issues much more easily.

In detail, we will go over the following in this chapter:

- **Configuring AWS:** Here, you set up a bucket and push your `mlruns` folders here to be stored on the cloud. These folders contain information about all of the runs associated with the experiments along with the logged models themselves. Next, you build a special Docker container as defined by MLFlow and push that to AWS ECR. SageMaker uses this container image to serve the MLFlow model.

- **Deploying a model to AWS SageMaker:** Here, you use the built-in MLFlow SageMaker module code to push a model to SageMaker. After SageMaker creates an endpoint, the model is hosted on here utilizing the docker image that you pushed earlier to the ECR.

- **Making predictions:** Once the model has finished deployment and is ready to serve, you use Boto3 to query the model and receive predictions.

- **Switching models:** MLFlow provides functionality that enables you to switch out a deployed model with a new one. SageMaker essentially updates the endpoint with the new model you are trying to deploy.

- **Removing the deployed model:** Finally, MLFlow lets you remove your deployed model altogether and delete the endpoint. This is important to do so that you don't incur the charges of leaving an endpoint running.

Also, it is important to note that AWS is actively being worked on, and functionality and operating procedures can change! What that means is that something that works now may not work later on.

However, MLFlow specifically provides support for SageMaker, so if something fundamental to how SageMaker runs changes in the future, MLFlow is likely to account for it in the next build.

In the absolute worst-case scenario where that doesn't happen, you can still run an MLFlow server and host it on AWS. You will still be able to deploy models and make inferences with them, and the overall functionality is still preserved. Instead of SageMaker directly hosting the model using an MLFlow container image, you would do something similar to the local model deployment experiment we did in Chapter 4, except you would connect to the server IP and port that the MLFlow server is hosted on.

We will explore how to do this with Google Cloud, as MLFlow does not support Google Cloud like it does SageMaker and Azure.

With that, let's get started!

Configuring AWS

Before you can actually push any model to SageMaker, you need to set up your Amazon workspace. You can push models from your local `mlruns` directory, similar to how you did local model deployment, but it is much more convenient and centralized to have all your runs be pushed to AWS and stored in a bucket. This way, all teams can access models that are stored here. In a sense, this can act as your "model registry," although it doesn't offer the same functionality as the model registry provided by MLFlow.

What MLFlow allows you to do is take specific runs and determine whether to stage that model to the development branch or to production. In this case, you can have buckets for each team, separated into development or production branches. It's a couple extra steps on top of MLFlow's model registry, but it would still allow you to enjoy the benefits of having a model registry.

In this case, you will simply be creating one bucket to host all of your MLFlow runs. From here, you will be picking a specific run and deploying to SageMaker. To keep it simple, you will once again use the scikit-learn logistic regression model that you trained as the model you are deploying.

So with that, create a simple bucket and name it something like `mlflow-sagemaker`. You can either create it through the AWS CLI or do so through the AWS console in your browser.

We will do the latter so that you can visually see what Amazon is really doing when a bucket is created.

Keep in mind that AWS is always working on its UI, so your screen may not look exactly like what is portrayed. That being said, you are still likely able to access S3 bucket storage services, so the core functionality should still be the same, despite the UI changes.

When you log into your portal, you should see something like Figure 5-1.

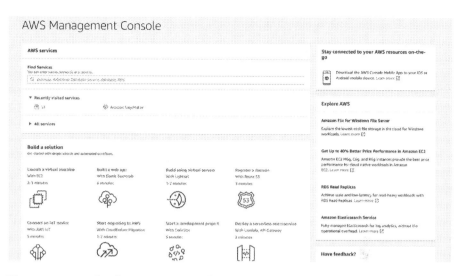

Figure 5-1. *The home screen of the AWS console. Keep in mind that yours is likely to look different to the one shown here*

As you can see, you can look up services with the search bar. Here, type S3 and click the result that states "S3" with the description "Scalable Storage in the Cloud."

You should go to a page that looks like Figure 5-2.

Figure 5-2. *What your screen might look like when you open the S3 bucket services module. We have greyed out the names of the buckets, but you can see string names here*

You should see a button that says Create Bucket. Click it and you will see something like Figure 5-3.

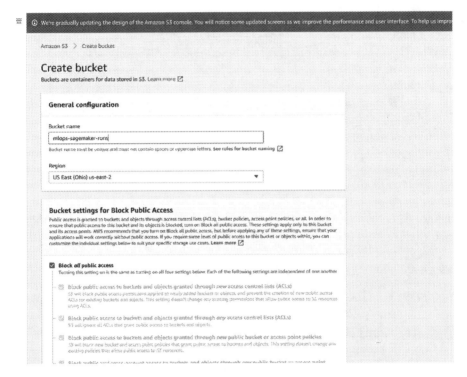

Figure 5-3. *This is how your bucket creation screen may look. In this case, you are just naming the bucket and aren't concerned with anything else*

We named our bucket `mlops-sagemaker-runs`. You don't have to worry about the rest of the options, so scroll down to the bottom and click Create Bucket. Once done, you should be able to see your bucket in the list of buckets.

From here, let's use a subprocess to sync the local `mlruns` directory to this bucket. What this does is upload the entire `mlruns` directory to your bucket, so that all of your runs are stored on the cloud.

First, collect the following attributes:

- s3_bucket_name: What is the name of the S3 bucket you are trying to push to?

- mlruns_directory: What is the location of the mlruns directory you're pushing to the bucket?

Based on that, run the following. We included the bucket name and mlruns directory in our case, so just replace them with your respective values.

```
import subprocess

s3_bucket_name = "mlops-sagemaker-runs"
mlruns_direc = "./mlruns/"

output = subprocess.run(["aws", "s3", "sync", "{}".
format(mlruns_direc), "s3://{}".format(s3_bucket_name)],
stdout=subprocess.PIPE, encoding='utf-8')
print(output.stdout)
print("\nSaved to bucket: ", s3_bucket_name)
```

After running that code, you should see something similar to Figure 5-4, letting you know that it has synchronized your local mlruns directory with the bucket. If you see no output, that means there's nothing new to push (if you are rerunning it). Ensure that the mlruns directory is in the same directory as this notebook; otherwise it won't be able to find it.

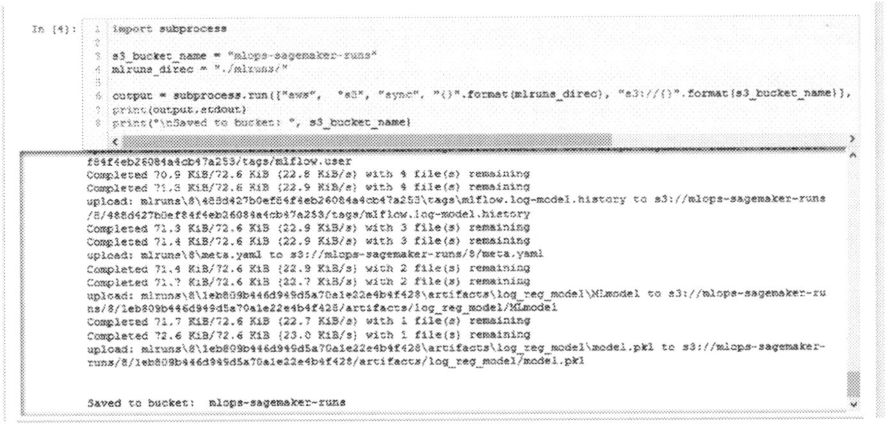

Figure 5-4. *This is what your output may look like when you are first syncing your mlruns directory with the bucket. Make sure that your mlruns directory is in the same directory as this notebook file*

Once this is done, you can proceed to building the container that SageMaker will use to host the model once you get to deployment. To do that, run the following command in your terminal:

```
mlflow sagemaker build-and-push-container
```

Again, this requires you to have your Amazon credentials configured.

You do not need to create a new docker image each time you use a new framework. This one image will be able to handle all your MLFlow models thanks to modularization. This is similar to the deployment pipeline we discussed in Chapter 3 from which you simply need to swap models in and out.

This step can take some time, so sit back, relax, and let it do its thing. You should see something like Figure 5-5.

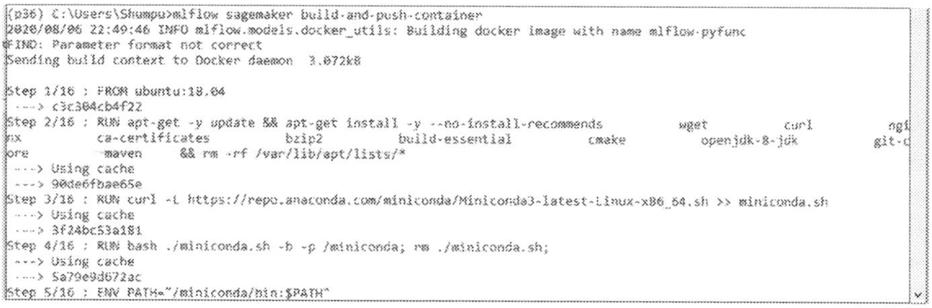

```
(p36) C:\Users\Shumpu>mlflow sagemaker build-and-push-container
2020/08/06 22:49:46 INFO mlflow.models.docker_utils: Building docker image with name mlflow-pyfunc
#FIND: Parameter format not correct
Sending build context to Docker daemon  3.072kB

Step 1/16 : FROM ubuntu:18.04
 ---> c3c304cb4f22
Step 2/16 : RUN apt-get -y update && apt-get install -y --no-install-recommends        wget        curl        ngi
nx        ca-certificates        bzip2        build-essential        cmake        openjdk-8-jdk        git-c
ore         maven      && rm -rf /var/lib/apt/lists/*
 ---> Using cache
 ---> 90de6fbae65e
Step 3/16 : RUN curl -L https://repo.anaconda.com/miniconda/Miniconda3-latest-Linux-x86_64.sh >> miniconda.sh
 ---> Using cache
 ---> 3f24bc53a181
Step 4/16 : RUN bash ./miniconda.sh -b -p /miniconda; rm ./miniconda.sh;
 ---> Using cache
 ---> 5a79e9d672ac
Step 5/16 : ENV PATH="/miniconda/bin:$PATH"
```

Figure 5-5. *Something similar to what you should see when you run the command to build the container*

Once this is finished, the console should output something like Figure 5-6.

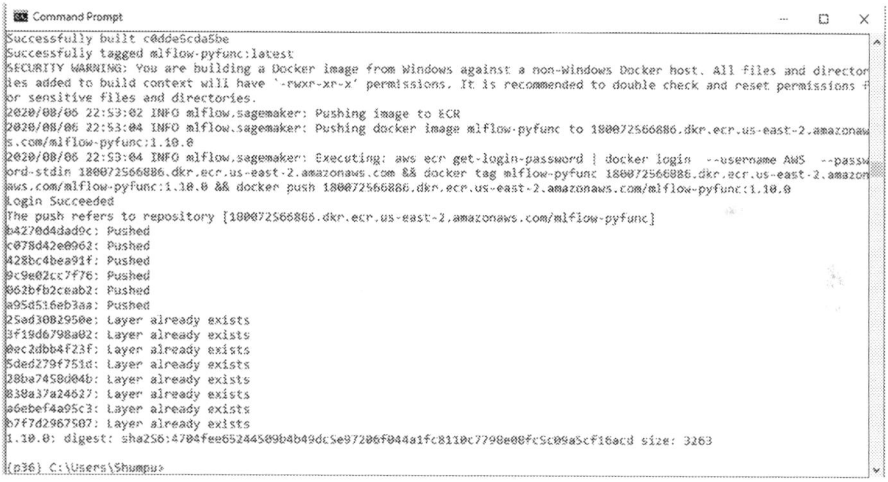

```
Command Prompt                                                          —    □    ×
Successfully built c0dde5cda5be
Successfully tagged mlflow-pyfunc:latest
SECURITY WARNING: You are building a Docker image from Windows against a non-Windows Docker host. All files and director
ies added to build context will have '-rwxr-xr-x' permissions. It is recommended to double check and reset permissions f
or sensitive files and directories.
2020/08/06 22:53:02 INFO mlflow.sagemaker: Pushing image to ECR
2020/08/06 22:53:04 INFO mlflow.sagemaker: Pushing docker image mlflow-pyfunc to 180072566886.dkr.ecr.us-east-2.amazonaw
s.com/mlflow-pyfunc:1.10.0
2020/08/06 22:53:04 INFO mlflow.sagemaker: Executing: aws ecr get-login-password | docker login  --username AWS  --passw
ord-stdin 180072566886.dkr.ecr.us-east-2.amazonaws.com && docker tag mlflow-pyfunc 180072566886.dkr.ecr.us-east-2.amazon
aws.com/mlflow-pyfunc:1.10.0 && docker push 180072566886.dkr.ecr.us-east-2.amazonaws.com/mlflow-pyfunc:1.10.0
Login Succeeded
The push refers to repository [180072566886.dkr.ecr.us-east-2.amazonaws.com/mlflow-pyfunc]
b4270d4dad9c: Pushed
c078d42e0962: Pushed
428bc4bea91f: Pushed
9c9e02cc7f76: Pushed
062bfb2ceab2: Pushed
a95d516eb3aa: Pushed
25ad3002950e: Layer already exists
3f19d6798a02: Layer already exists
0ec2dbb4f23f: Layer already exists
5ded279f751d: Layer already exists
28ba7458d04b: Layer already exists
838a37a24627: Layer already exists
a6ebef4a95c3: Layer already exists
b7f7d2967587: Layer already exists
1.10.0: digest: sha256:4704fee65244509b4b49dc5e97206f044a1fc8110c7798e08fc5c09a5cf16acd size: 3263

(p36) C:\Users\Shumpu>
```

Figure 5-6. *What you should see when the docker container image has successfully been built and pushed to Amazon ECR*

Now, you should be able to see a new container in the portal when you navigate to Amazon ECR.

From your home console, navigate to Amazon ECR, and verify you see something called `mlflow-pyfunc`. You should see something like Figure 5-7, confirming that the docker image has successfully been pushed to AWS ECR.

Figure 5-7. *After running the command, you should be able to see your container in the ECR repository list*

With that, you have set up everything related to MLFlow functionality that you need in your AWS console in order to deploy your models to SageMaker.

Let's now look at deploying one of the models.

Deploying a Model to AWS SageMaker

To deploy a model to SageMaker, you need to gather the following information:

- `app_name`
- `model_uri`
- `execution_role`
- `region`
- `image_ecr_url`

The execution role refers to the Identity and Access Management (IAM) role, which you can find by searching for "IAM" in the console. Once you have created or selected an execution role (make sure it can access S3 and can perform get, put, delete, and list operations on it), copy the entire value that exists there.

As for the specific policy that this role should follow, refer to Figure 5-8 to see how our IAM execution role is set up.

As for the execution role ARN number, you should see something like Figure 5-8.

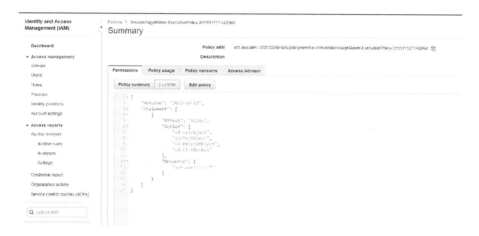

Figure 5-8. *In the IAM tab, under policies, select (or create) the role you are going to use to execute the deployment process. There, you should be able to see the specific Policy ARN value, which you must copy and keep track of*

Make sure you have the Policy ARN value copied down. AWS lets you copy it to the clipboard if you click the little clipboard symbol next to the policy.

To find the `image_ecr_url` value, go back to the ECR and look for something like Figure 5-7. Now click it to see something like Figure 5-9.

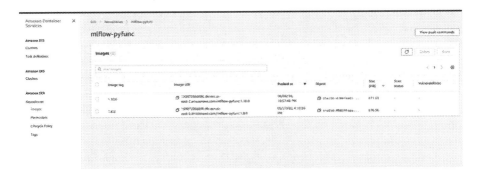

Figure 5-9. *The Image URI is the value you want to copy*

Copy the value where it says Image URI, except for the version you want. We are running MLFlow version 1.10.0, so copy the value for that one.

Next, find the specific run that you want to deploy. Go to your list of S3 buckets and click the one you created, which should be titled `mlops-sagemaker-runs`.

In here, navigate until you see the folder with several runs displayed. We picked the top run. Refer to Figure 5-10.

Figure 5-10. *Look at your bucket to find the run you want to deploy. (These runs all have the same performance metrics, so it does not matter which one we pick. If it did, we could look at it through the MLFlow UI (ensuring the terminal is in the same directory as the same mlruns directory we pushed) and select the best run.) Also, remember to take note of the experiment ID and the name of the model you logged. You should be able to find it if you click the run ID and then artifacts. For our case, it is log_reg_model*

With all that information gathered, let's proceed to the deployment. Run the following:

```
import boto3
import mlflow.sagemaker as mfs
import json

app_name = "mlops-sagemaker"
execution_role_arn = "arn:aws:iam::180072566886:role/
service-role/AmazonSageMaker-ExecutionRole-20181112T142060"
image_ecr_url = "180072566886.dkr.ecr.us-east-2.amazonaws.com/
mlflow-pyfunc:1.10.0"
region = "us-east-2"

s3_bucket_name = "mlops-sagemaker-runs"
experiment_id = "8"
run_id = "1eb809b446d949d5a70a1e22e4b4f428"
model_name = "log_reg_model"

model_uri = "s3://{}/{}/{}/artifacts/{}/".format
(s3_bucket_name, experiment_id, run_id, model_name)
```

This will set up all of the parameters that you will use to run the deployment code.

Finally, let's get on to the actual deployment code:

```
mfs.deploy(app_name=app_name,
           model_uri=model_uri,
           execution_role_arn=execution_role_arn,
           region_name=region,
           image_url=image_ecr_url,
           mode=mfs.DEPLOYMENT_MODE_CREATE)
```

You should see something like Figure 5-11.

Figure 5-11. *You should see something like this when you are attempting to deploy the model. Don't worry if it takes its time*

This step can take a while. If you want to check on the status of your SageMaker endpoint, open up the portal and search for and navigate to SageMaker. There should be a section for Endpoints where you can see all of the SageMaker endpoints that exist. You should see your current endpoint with the status of "creating," as in Figure 5-12.

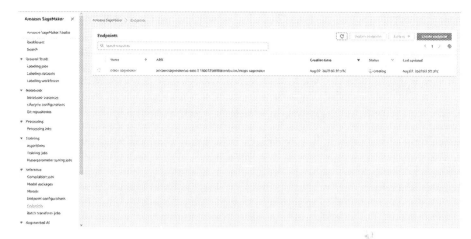

Figure 5-12. *What you should see in the Endpoints section of Amazon SageMaker. Once it has finished creating the endpoint, you should see it update the status to "InService."*

Once this endpoint is successfully created, which you will know when you see the status update to "InService," you can now move on to making predictions.

Making Predictions

Making predictions is simple. All you need is the name of the endpoint and the functionality that boto3 provides in order for the model to be queried. Let's define a function to query the model:

```
def query(input_json):

        client = boto3.session.Session().client
        ("sagemaker-runtime", region)

        response = client.invoke_endpoint(
            EndpointName=app_name,
            Body=input_json,
            ContentType='application/json; format=pandas-split',
        )
```

```
        preds = response['Body'].read().decode("ascii")
        preds = json.loads(preds)
        return preds
```

Now, let's load your data, process it, and scale it just like you did for the local model deployment example. Make sure that the folder data exists, ensuring that creditcard.csv exists within it. Run the following:

```python
import pandas as pd
import mlflow
import mlflow.sklearn

import seaborn as sns

import matplotlib.pyplot as plt

from sklearn.preprocessing import StandardScaler
from sklearn.model_selection import train_test_split
from sklearn.metrics import roc_auc_score, accuracy_score,
confusion_matrix

import numpy as np

df = pd.read_csv("data/creditcard.csv")
```

Once the import statements and the data frame has been loaded, run the following:

```python
normal = df[df.Class == 0].sample(frac=0.5, random_state=2020).
reset_index(drop=True)
anomaly = df[df.Class == 1]

normal_train, normal_test = train_test_split(normal,
test_size = 0.2, random_state = 2020)
anomaly_train, anomaly_test = train_test_split(anomaly,
test_size = 0.2, random_state = 2020)
```

```
scaler = StandardScaler()
scaler.fit(pd.concat((normal, anomaly)).drop(["Time",
"Class"], axis=1))
```

Once this is all finished, run the following to ensure that the model is actually making predictions:

```
scaled_selection = scaler.transform(df.iloc[:80].drop
(["Time", "Class"], axis=1))
input_json = pd.DataFrame
(scaled_selection).to_json(orient="split")

pd.DataFrame(query(input_json)).T
```

You should see an output like Figure 5-13.

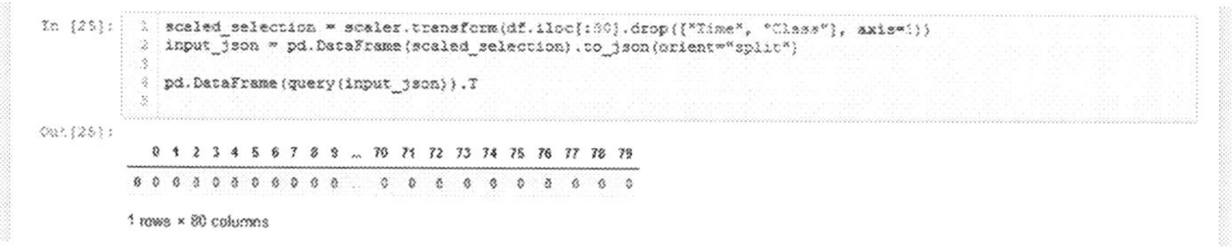

Figure 5-13. *Querying the deployed model with the scaled data representing the first 80 rows of the data frame and getting a response back*

Figure 5-13 shows a successful query of the model while it is hosted on a SageMaker endpoint and the predictions received as a response.

Let's run the batch query script with some modifications:

```
test = pd.concat((normal.iloc[:1900], anomaly.iloc[:100]))
true = test.Class
test = scaler.transform(test.drop(["Time", "Class"], axis=1))
preds = []
```

```
batch_size = 80
for f in range(25):
    print(f"Batch {f}", end=" - ")

    sample = pd.DataFrame(test[f*batch_size:(f+1)*batch_size]).
    to_json(orient="split")

    output = query(sample)
    resp = pd.DataFrame([output])
    preds = np.concatenate((preds, resp.values[0]))

    print("Completed")

eval_acc = accuracy_score(true, preds)
eval_auc = roc_auc_score(true, preds)

print("Eval Acc", eval_acc)
print("Eval AUC", eval_auc)
```

Once finished, you should see something like Figure 5-14.

```
In [36]:   1  test = pd.concat([normal.iloc[:1900], anomaly.iloc[:130]])
           2  true = test.Class
           3  test = scaler.transform(test.drop(["Time", "Class"], axis=1))
           4  preds = []
           5
           6  batch_size = 80
           7  for f in range(25):
           8      print(f"Batch {f}", end=" - ")
           9
          10      sample = pd.DataFrame(test[f*batch_size:(f+1)*batch_size]).to_json(orient="split")
          11
          12      output = query(sample)
          13      resp = pd.DataFrame([output])
          14      preds = np.concatenate([preds, resp.values[0]])
          15
          16      print("Completed")
          17
          18  eval_acc = accuracy_score(true, preds)
          19  eval_auc = roc_auc_score(true, preds)
          20
          21  print("Eval Acc", eval_acc)
          22  print("Eval AUC", eval_auc)

Batch 0 - Completed
Batch 1 - Completed
Batch 2 - Completed
Batch 3 - Completed
Batch 4 - Completed
Batch 5 - Completed
Batch 6 - Completed
Batch 7 - Completed
Batch 8 - Completed
Batch 9 - Completed
Batch 10 - Completed
Batch 11 - Completed
Batch 12 - Completed
Batch 13 - Completed
Batch 14 - Completed
Batch 15 - Completed
Batch 16 - Completed
Batch 17 - Completed
Batch 18 - Completed
Batch 19 - Completed
Batch 20 - Completed
Batch 21 - Completed
Batch 22 - Completed
Batch 23 - Completed
Batch 24 - Completed
Eval Acc 0.9915
Eval AUC 0.915
```

Figure 5-14. *Output of the batch querying script. You included a mix of 100 anomalies with 1900 normal points so that you can get a better idea of how the model performs against anomalies as well. Otherwise, you would have gotten a handful of anomalies*

All this is great, but what do you do when you want to switch the model that is deployed? Well, SageMaker allows you to update the endpoint and switch to a new model. Let's look at how to do this.

Switching Models

Perhaps you want to update your model, or you have no more use for the current model and its prediction services so you want to replace it without having to delete and create a new endpoint. In this case, you can simply update the endpoint and swap out the model that is currently hosted on there. To do so, you only need to collect the new model_uri.

This time, the model_uri refers to the URI of the new model that you want to deploy. In your case, you are selecting the second run of the three runs you uploaded to your bucket. Everything else remains the same, so you only have to get a new model_uri.

Now, run the following, replacing the run_id value with your chosen run_id:

```
new_run_id = "3862eb3bd89b43e8ace610c521d974e6"

new_model_uri = "s3://{}/{}/{}/artifacts/{}/".format
(s3_bucket_name, experiment_id, new_run_id, model_name)
```

Now that you have run this, run the following code to update the model:

```
mfs.deploy(app_name=app_name,
           model_uri=new_model_uri,
           execution_role_arn=execution_role_arn,
           region_name=region,
           image_url=image_ecr_url,
           mode=mfs.DEPLOYMENT_MODE_REPLACE)
```

You will find that this function looks quite similar to the one you used to deploy the model. The only parameter that differs is the mode, as you are now doing mfs.DEPLOYMENT_MODE_REPLACE instead of mfs. DEPLOYMENT_MODE_CREATE.

Refer to Figure 5-15 to see what the output should look like. Note that this also can take some time to finish.

```
In [*]:  1  mfs.deploy(app_name=app_name,
         2      model_uri=new_model_uri,
         3      execution_role_arn=execution_role_arn,
         4      region_name=region,
         5      image_url=image_ecr_url,
         6      mode=mfs.DEPLOYMENT_MODE_REPLACE)
```

```
2020/08/07 00:19:31 INFO mlflow.sagemaker: Using the python_function flavor for deployment!
2020/08/07 00:19:32 INFO mlflow.sagemaker: No model data bucket specified, using the default bucket
2020/08/07 00:19:33 INFO mlflow.sagemaker: Default bucket 'mlflow-sagemaker-us-east-2-180072566886' already ex
ists. Skipping creation.
2020/08/07 00:19:34 INFO mlflow.sagemaker: tag response: {'ResponseMetadata': {'RequestId': '19E60C275A0F8688
', 'HostId': 'm29MKVESolavY1FuIqPYjO0yCx1BFwgoQfmBv7OntEYz90XjRfmaQIq06Zf16+m4YZJQ/1N5n+s=', 'HTTPStatusCode':
200, 'HTTPHeaders': {'x-amz-id-2': 'm29MKVESolavY1FuIqPYjO0yCx1BFwgoQfmBv7OntEYz90XjRfmaQIq06Zf16+m4YZJQ/1N5n+
s=', 'x-amz-request-id': '19E60C275A0F8688', 'date': 'Fri, 07 Aug 2020 04:19:34 GMT', 'content-length': '0', '
server': 'AmazonS3'}, 'RetryAttempts': 0}}
2020/08/07 00:19:34 INFO mlflow.sagemaker: Found active endpoint with arn: arn:aws:sagemaker:us-east-2:1800725
66886:endpoint/mlops-sagemaker. Updating...
2020/08/07 00:19:34 INFO mlflow.sagemaker: Created new model with arn: arn:aws:sagemaker:us-east-2:18007256688
6:model/mlops-sagemaker-model-o5kmdkSbtxabtabice-0tbrq
2020/08/07 00:19:34 INFO mlflow.sagemaker: Created new endpoint configuration with arn: arn:aws:sagemaker:us-e
ast-2:180072566886:endpoint-config/mlops-sagemaker-config-npdknzrpskwicek40ulbvq
2020/08/07 00:19:35 INFO mlflow.sagemaker: Updated endpoint with new configuration!
2020/08/07 00:19:35 INFO mlflow.sagemaker: Waiting for the deployment operation to complete...
2020/08/07 00:19:35 INFO mlflow.sagemaker: The operation is still in progress.
2020/08/07 00:19:55 INFO mlflow.sagemaker: The update operation is still in progress. Current endpoint status:
"Updating"
2020/08/07 00:20:16 INFO mlflow.sagemaker: The update operation is still in progress. Current endpoint status:
"Updating"
2020/08/07 00:20:36 INFO mlflow.sagemaker: The update operation is still in progress. Current endpoint status:
"Updating"
```

Figure 5-15. *This is what your output should look like after running the update code*

While this is running, you can check on the endpoint in your portal to see that it is now updating. Refer to Figure 5-16 to see this.

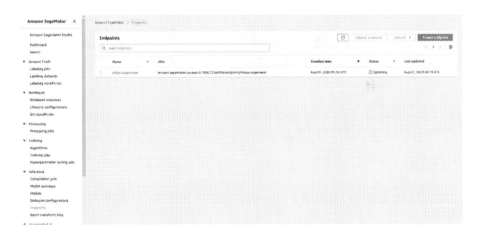

Figure 5-16. *The endpoint is now updating. Once finished, it should show "InService" just like when the endpoint was being created*

Once it finishes running, you can query this model again using the same function. You don't have to modify the batch script either.

Now that you know how to update the endpoint with a new model, we will look at how you can remove the endpoint and the deployed model.

Removing Deployed Model

Perhaps you have multiple endpoints each with a different model hosted, and you no longer want to keep an endpoint running because of the cost. To delete an endpoint, you only need the following information:

- app_name

- region

With that information defined, which it already should be, you can simply run the following:

mfs.delete(app_name=app_name,region_name=region)

You should see it output something like Figure 5-17. This process finishes quite quickly.

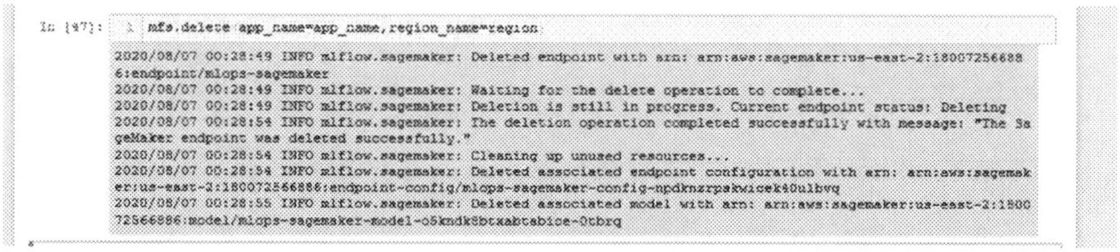

Figure 5-17. *The output of the deletion command*

You can go check the endpoint in the portal as well, and it should show something like Figure 5-18.

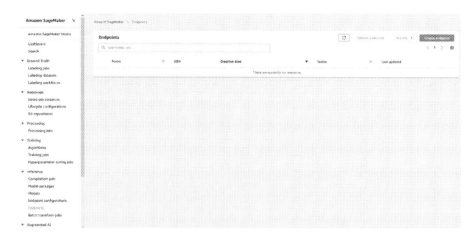

Figure 5-18. *SageMaker endpoint resources after the deletion. There should be nothing here if the deletion process went successfully*

As you can see, the endpoint is now completely gone.

One thing to note is that you should make sure you don't accidentally leave any resources running because the costs can certainly stack up over time and put a dent in your wallet. For services like SageMaker endpoints, you are charged by the hour, so be sure to delete them once you're done with them.

As for the S3 bucket and the ECR container, those are a one-time charge that only bill for data transfer.

With that, you now know how to operationalize your MLFlow model with AWS SageMaker.

Summary

MLFlow provides explicit AWS SageMaker support in its operationalization code. And so we covered how to upload your runs to an S3 bucket and how to create and push an MLFlow Docker container image for AWS SageMaker to use when operationalizing your models. We also covered how to deploy your model on an endpoint, query it, update the endpoint with a new model, and delete the endpoint. Hopefully now you now know how to operationalize your machine learning models with MLFlow and AWS SageMaker.

In the next chapter, we will look at how you can operationalize your MLFlow models with Microsoft Azure.

CHAPTER 6

Deploying in Azure

In this chapter, we will cover how you can use Microsoft Azure to operationalize your MLFlow models. In particular, we will look at how you can also utilize Azure's built-in functionality to deploy a model to a development branch and to a production branch, along with how you can query the models once deployed.

Introduction

In the previous chapter, we went over how to deploy your models to Amazon SageMaker, manage them through update or delete events, and query them. Now, we will shift our focus to show how you can operationalize your MLFlow models using Microsoft Azure.

Before you begin, here is an **important prerequisites**:

- Install azureml-sdk in your Python environment.

Just like with AWS, Microsoft Azure is constantly being worked on and updated. Since MLFlow supports Microsoft Azure, you should be able to utilize MLFlow to operationalize your models. Any new functionality is sure to be documented by MLFlow, and in the absolute worst-case scenario, you should still be able to host a server on Azure and maintain your MLOps functionality that way.

Again, we will explore how to do this in the next chapter when we look at how to operationalize your MLFlow models with the Google Cloud API.

© Sridhar Alla, Suman Kalyan Adari 2021
S. Alla and S. K. Adari, *Beginning MLOps with MLFlow*,
https://doi.org/10.1007/978-1-4842-6549-9_6

In detail, we will go over the following in this chapter:

- **Configuring Azure:** Here, you basically use MLFlow's functionality to build a container image for the model to be hosted in. Then, you push it to Azure's Azure Container Instances (ACI), similar to how you pushed an image to the Amazon AWS Elastic Container Registry (ECR).

- **Deploying a model to Azure (dev stage):** Here, you use built-in azureml-sdk module code to push a model to Azure. However, this is a development stage deployment, so this model is not production-ready since its computational resources are limited.

- **Making predictions:** Once the model has finished deployment, it is ready to be queried. This is done through an HTTP request. This is how you can verify that your model works once hosted on the cloud since it's in the development stage.

- **Deploying to production:** Here, you utilize MLFlow Azure module code to deploy the model to production by creating a container instance (or any other deployment configuration provided, like Azure Kubernetes Service).

- **Making predictions:** Similar to how you query the model in the dev stage, you query the model once it has been deployed to the production stage and run the batch query script from the previous chapter.

- **Switching models:** MLFlow does not provide explicit functionality to switch your models, so you must delete the service and recreate it with another model run.

- **Removing the deployed model:** Finally, you undo every deployment that you did and remove all resources. That is, you delete both the development and production branch services as well as the container registries and any additional services created once you are done.

With that, let's get started!

Configuring Azure

Before you can start using Azure's functionality to operationalize your models, you must first create or connect to an existing Azure workspace. You can do this either through code or the UI in a browser.

In your case, you will open up the portal in the browser and learn how to create a workspace. Refer to Figure 6-1.

Figure 6-1. *An example of the Microsoft Azure portal home screen*

Next, click the Create a resource option and search for "Machine Learning." You should see something like Figure 6-2.

Figure 6-2. *An example of the service "Machine Learning" provided by Azure. You want to create a workspace within this service, so click the Create button*

Click the Create button. You should see something like Figure 6-3. (We filled the fields with our own parameters.)

Your subscription might differ from ours. For the resource group, we created a new one titled `azure-mlops`.

The fields you completed in Figure 6-3 are enough to create your workspace. Next, click the Review + create option and click Create once Azure states that the validation procedure has been passed and allows you to click Create.

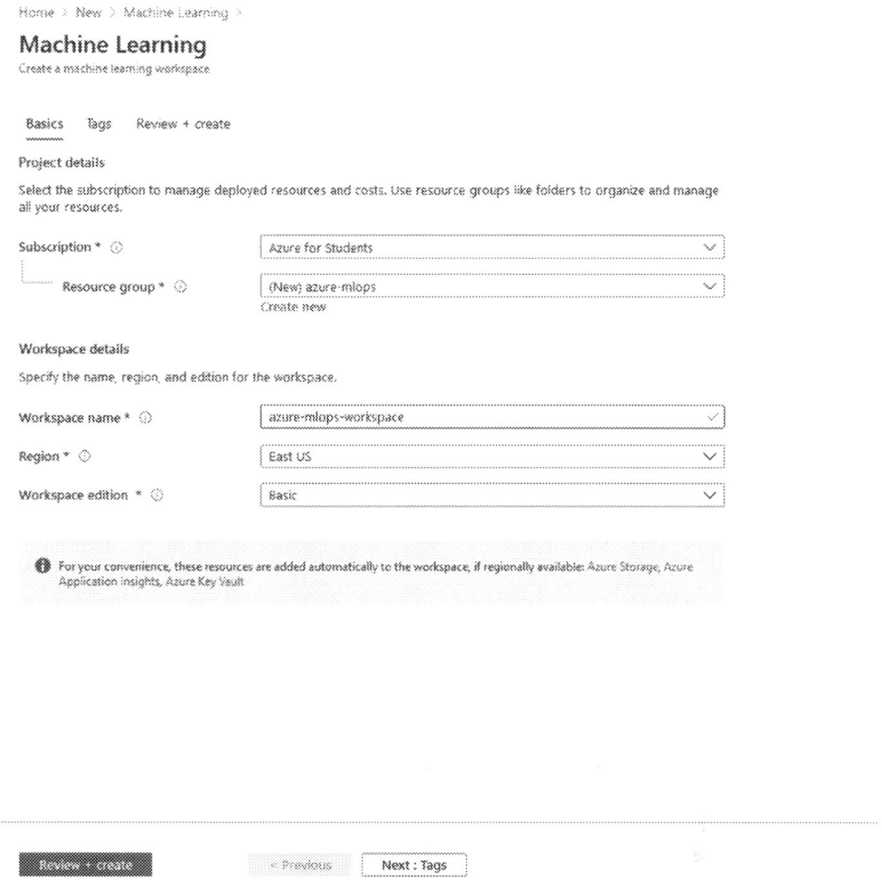

Home > New > Machine Learning >

Machine Learning
Create a machine learning workspace

Basics   Tags   Review + create

Project details

Select the subscription to manage deployed resources and costs. Use resource groups like folders to organize and manage all your resources.

Subscription * ⓘ

Azure for Students   ⌄

‎   Resource group * ⓘ

(New) azure-mlops   ⌄
Create new

Workspace details

Specify the name, region, and edition for the workspace.

Workspace name * ⓘ

azure-mlops-workspace   ⌄

Region * ⓘ

East US   ⌄

Workspace edition * ⓘ

Basic   ⌄

ⓘ For your convenience, these resources are added automatically to the workspace, if regionally available: Azure Storage, Azure Application Insights, Azure Key Vault

Review + create   < Previous   Next : Tags

Figure 6-3. *Workspace creation UI (we filled in the fields with our own parameters)*

This will take some time to deploy. Once the workspace has been created, go back to the home portal and click the All resources option. You should see something like Figure 6-4.

Click your workspace, which should have an image of a chemical beaker next to it.

In this overview, you will see several parameters associated with this workspace. Make sure to keep track of the following attributes of the workspace so that you can connect to it in the code:

257

- workspace_name (azure-mlops-workspace)

- subscription (The value where it says Subscription-ID)

- resource_group (azure-mlops)

- location (East-US)

Figure 6-4. *You might see something like this when you look at the All resources option*

Refer to Figure 6-5.

Figure 6-5. *You should see something like this for your own workspace. Here we've censored potentially sensitive fields, but you should be able to see your own unique subscription ID on your screen. This is the value you want to use*

Now that you have that, run the following to create/connect to your own workspace:

```python
import azureml
from azureml.core import Workspace

workspace_name = "MLOps-Azure"
workspace_location="East US"
resource_group = "mlflow_azure"
subscription_id = "xxxxxxxx-xxxx-xxxx-xxxx-xxxxxxxxxxxx"

workspace = Workspace.create(name = workspace_name,
                             location = workspace_location,
                             resource_group = resource_group,
                             subscription_id = subscription_id,
                             exist_ok=True)
```

If you have successfully connected to your workspace, the cell should run without any issues.

Next, you must build the MLFlow container image to be used by Azure. Here, you also specify the run of the model you are trying to deploy.

In the case of Amazon SageMaker, you were able to reference runs from your local machine or runs from an S3 bucket. You can do the same thing for Azure, except using Azure's storage entities called blobs.

Either way, you need the **run ID** of the model you are deploying and the **artifact scheme** that the model is logged in. For the models you stored in Amazon S3 buckets, you used the scheme s3:/, but this time you will just use a run locally. If you'd like, you can still use your Amazon S3 bucket or Google Cloud buckets. Where you store your run does not matter.

Run the following, replacing the values with your specific run and storage scheme:

```python
run_id = "1eb809b446d949d5a70a1e22e4b4f428"
model_name = "log_reg_model"
model_uri = f"runs:/{run_id}/{model_name}"
```

The model name should be the same in your case unless you changed it. Since we are using local runs, we have a URI starting with `runs:/`. Again, change this to whatever is appropriate in your case.

Finally, with all that information set, let's create the container image:

```
import mlflow.azureml

model_image, azure_model = mlflow.azureml.build_image
                            (model_uri=model_uri,
                            workspace=workspace,
                            model_name="sklearn_logreg_dev",
                            image_name="model",
                            description="SkLearn LogReg Model
                            for Anomaly Detection",
                            synchronous=False)
```

You should see something like Figure 6-6. You may or may not see the warning messages depending on your version of MLFlow.

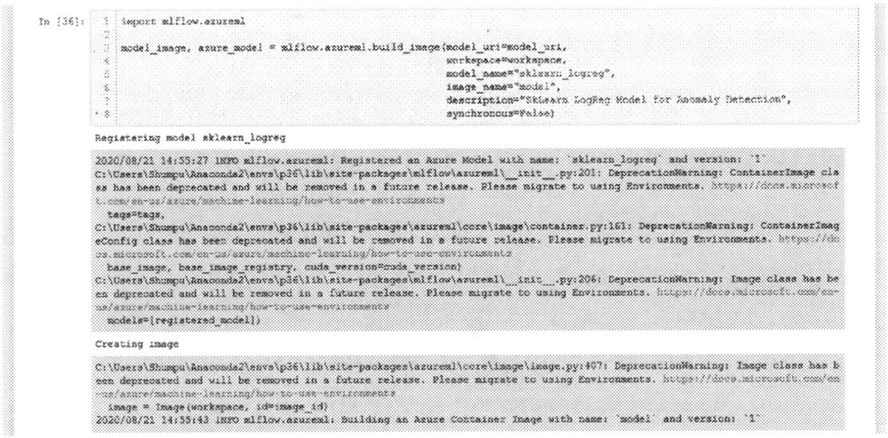

Figure 6-6. *Building and pushing the container to Azure's container registry. Ignore the warning messages for now. You might not see these messages in the future. Since this is code created and maintained by MLFlow, it is likely that they will provide support for whatever new functionality Azure pushes*

Next, run the following to check the status of the container:

```
model_image.wait_for_creation(show_output=True)
```

You should see something like Figure 6-7.

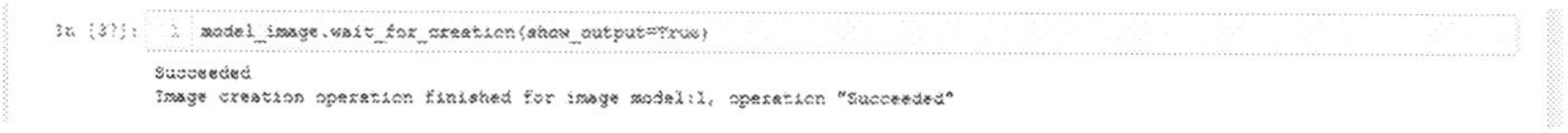

Figure 6-7. *Checking the output of the progress in the image creation operation*

Once the image has been created, you can now deploy your model.

Deploying to Azure (Dev Stage)

One interesting bit of functionality that Azure provides is the ACI webservice. This webservice is specifically used for the purposes of debugging or testing some model under development, hence why it is suitable for use in the development stage.

You are going to deploy an ACI webservice instance based on the model image you just created.

Run the following:

```
from azureml.core.webservice import AciWebservice, Webservice

aci_service_name = "sklearn-model-dev"
aci_service_config = AciWebservice.deploy_configuration()

aci_service = Webservice.deploy_from_image
            (name=aci_service_name,
            image=model_image,
            deployment_config=aci_service_config,
            workspace=workspace)
```

You should see something like Figure 6-8.

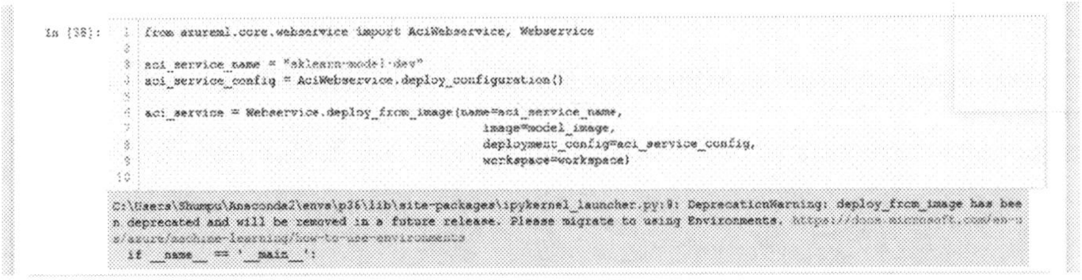

Figure 6-8. *The output of creating the ACI service. It seems that this function may be removed in the future, but for now this is one way to access the ACI service and deploy the model*

This exact way of starting the service may be deprecated in the near future in favor of Environments. For the time being, you should still be able to start an ACI service in this manner, but the important thing to know is that there is a web service specifically tailored for development stage testing.

Now run the following to check the progress:

```
aci_service.wait_for_deployment(show_output=True)
```

You should see something like Figure 6-9.

Figure 6-9. *The output you should see from checking if the deployment has succeeded*

Before making your predictions, let's first verify that you can reach your service:

```
aci_service.scoring_uri
```

You should see something like Figure 6-10. If not, try going into your resources in the portal to verify that a new container exists with the name sklearn-model-dev. If not, try rerunning the cells in the same order. It should display some URI this time.

You should see something like Figure 6-10.

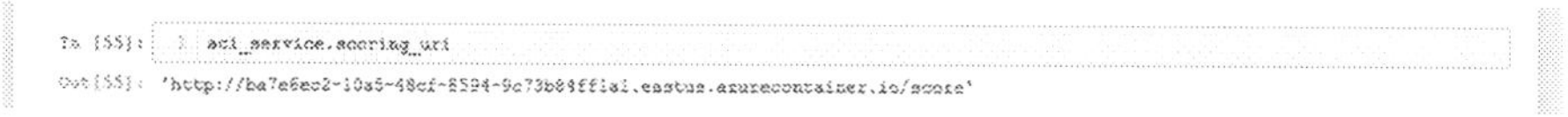

Figure 6-10. *The scoring URI is displayed, indicating that you can connect to it and make predictions*

You can now make predictions with this model.

Making Predictions

Now you need to acquire some data to predict with.

Just like before, you will be loading the credit card dataset, preprocessing it, and setting aside a small batch that you will query the model with. Run the following blocks of code, and make sure you have the folder named data in this directory with creditcard.csv in it:

```
import pandas as pd
import mlflow
import mlflow.sklearn

import seaborn as sns

import matplotlib.pyplot as plt

from sklearn.preprocessing import StandardScaler
from sklearn.model_selection import train_test_split
from sklearn.metrics import roc_auc_score, accuracy_score,
confusion_matrix
```

```
import numpy as np

import subprocess
import json

df = pd.read_csv("data/creditcard.csv")
```

Once you have loaded all the modules and have loaded the data, run the following:

```
normal = df[df.Class == 0].sample(frac=0.5, random_state=2020).
reset_index(drop=True)
anomaly = df[df.Class == 1]

normal_train, normal_test = train_test_split(normal, test_size
= 0.2, random_state = 2020)
anomaly_train, anomaly_test = train_test_split(anomaly,
test_size = 0.2, random_state = 2020)

scaler = StandardScaler()
scaler.fit(pd.concat((normal, anomaly)).drop(["Time", "Class"],
axis=1))
```

In cells, the above two blocks of code should look like Figure 6-11.

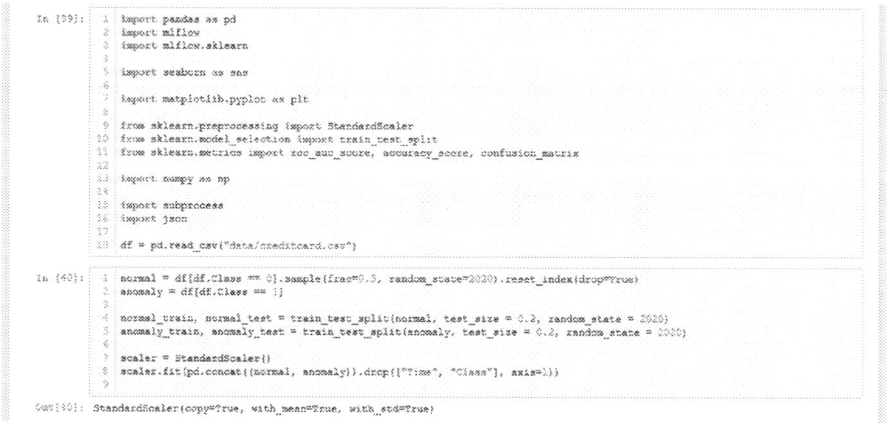

Figure 6-11. *The import statements and data processing code. You also define the scaler here and fit it to the data, just as you did when originally training these models*

Once you are all done with preparing the data, let's define a function to help you query the deployed model:

```python
import requests
import json

def query(scoring_uri, inputs):

    headers = {
    "Content-Type": "application/json",
    }

    response = requests.post(scoring_uri, data=inputs,
    headers=headers)
    preds = json.loads(response.text)
    return preds
```

Now you can select a few points and make a prediction:

```
data_selection = df.iloc[:80].drop(["Time", "Class"], axis=1)

input_json = pd.DataFrame(scaler.transform(data_selection)).
to_json(orient="split")

preds = query(scoring_uri=aci_service.scoring_uri,
inputs=input_json)
pd.DataFrame(preds).T
```

Together, you should see something like Figure 6-12.

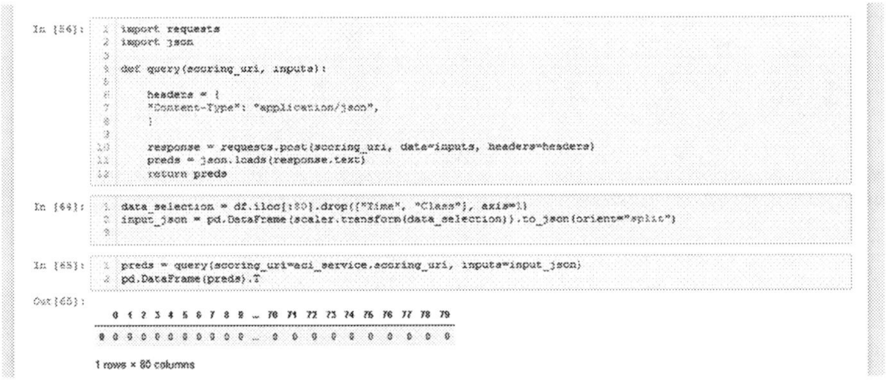

Figure 6-12. *Querying the model deployed on an ACI webservice with some sample data and receiving a response*

As you can see, the model has returned predictions that look correct (thanks to the scaling).

Now that you know how to deploy to a development branch, let's look at how you can deploy the model to production using built-in MLFlow functionality.

Deploying to Production

MLFlow provides Azure support and helps us deploy our models directly, using a container instance by default.

Let's get straight into it. Run the following, replacing the names with anything else preferred:

```
azure_service, azure_model = mlflow.azureml.deploy(model_uri,
                    workspace,
                    service_name="sklearn-logreg",
                    model_name="log-reg-model",
                    synchronous=True)
```

It's worth mentioning that you can deploy to a specific web service. By default, MLFlow will host the model on a container instance, but you can specify a computer cluster. To learn more, refer to the documentation here:

www.mlflow.org/docs/latest/python_api/mlflow.azureml.html.

Once the code finishes running, which can take some time, you can also check to see if the URI can be printed:

```
azure_service.scoring_uri
```

Together, you should see something like Figure 6-13.

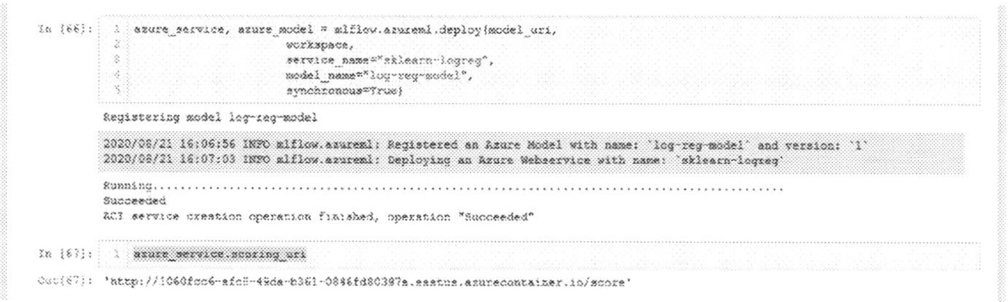

Figure 6-13. *Successfully creating the endpoint and verifying that the service has a URI*

Now that you have successfully deployed your model, let's move on to making predictions.

Making Predictions

Now that you have your model deployed, let's run your code to make predictions.

First of all, let's run the following to make sure that you are receiving predictions. You should already have defined input_json:

```
preds = query(scoring_uri=azure_service.scoring_uri,
inputs=input_json)
pd.DataFrame(preds).T
```

You should now see something like Figure 6-14.

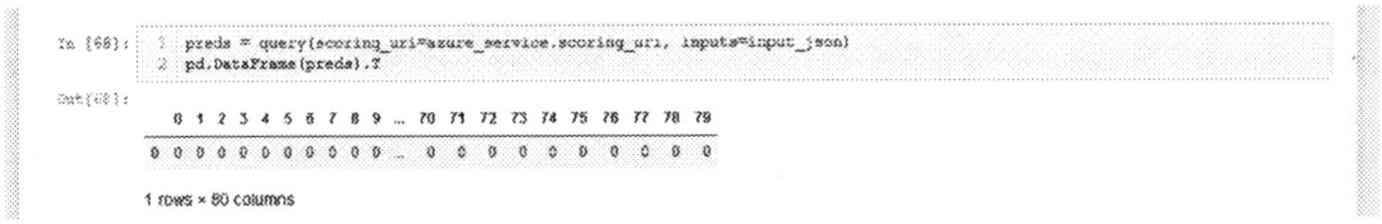

Figure 6-14. *Querying the deployed model with your batch of scaled data to ensure it works*

Now, let's run your batch querying script:

```
test = pd.concat((normal.iloc[:1900], anomaly.iloc[:100]))
true = test.Class
test = scaler.transform(test.drop(["Time", "Class"], axis=1))
preds = []
```

```
batch_size = 80
for f in range(25):
    print(f"Batch {f}", end=" - ")

    sample = pd.DataFrame(test[f*batch_size:(f+1)*batch_size]).
    to_json(orient="split")

    output = query(scoring_uri=azure_service.scoring_uri,
    inputs=sample)
    resp = pd.DataFrame([output])
    preds = np.concatenate((preds, resp.values[0]))

    print("Completed")

eval_acc = accuracy_score(true, preds)
eval_auc = roc_auc_score(true, preds)

print("Eval Acc", eval_acc)
print("Eval AUC", eval_auc)
```

Once finished, you should see something like Figure 6-15.

```
In [69]:   1   test = pd.concat((normal.iloc[:1900], anomaly.iloc[:100]))
           2   true = test.Class
           3   test = scaler.transform(test.drop([["Time", "Class"], axis=1))
           4   preds = []
           5
           6   batch_size = 80
           7   for f in range(25):
           8       print(f"Batch {f}", end=" - ")
           9
          10       sample = pd.DataFrame(test[f*batch_size:(f+1)*batch_size]).to_json(orient="split")
          11
          12       output = query(scoring_uri=azure_service.scoring_uri, inputs=sample)
          13       resp = pd.DataFrame([output])
          14       preds = np.concatenate((preds, resp.values[0]))
          15
          16       print("Completed")
          17
          18   eval_acc = accuracy_score(true, preds)
          19   eval_auc = roc_auc_score(true, preds)
          20
          21   print("Eval Acc", eval_acc)
          22   print("Eval AUC", eval_auc)

Batch 0 - Completed
Batch 1 - Completed
Batch 2 - Completed
Batch 3 - Completed
Batch 4 - Completed
Batch 5 - Completed
Batch 6 - Completed
Batch 7 - Completed
Batch 8 - Completed
Batch 9 - Completed
Batch 10 - Completed
Batch 11 - Completed
Batch 12 - Completed
Batch 13 - Completed
Batch 14 - Completed
Batch 15 - Completed
Batch 16 - Completed
Batch 17 - Completed
Batch 18 - Completed
Batch 19 - Completed
Batch 20 - Completed
Batch 21 - Completed
Batch 22 - Completed
Batch 23 - Completed
Batch 24 - Completed
Eval Acc 0.9915
Eval AUC 0.915
```

Figure 6-15.   The results of running the batch querying script. This effectively made predictions on 2,000 data points

With that, you now know how to query your deployed model and make predictions with it. This should be the same procedure if you've opted to deploy to a specific compute cluster with, for example, Azure Kubernetes Service.

Cleaning Up

Unfortunately, there does not seem to be any specific functionality to **update** the service with a new model. The procedure seems to be to delete the service and create a new service with another model URI.

So, with that, let's now look at how you can remove all the services you just created.

Run the following:

```
aci_service.delete()
azure_service.delete()
```

Refer to Figure 6-16.

Figure 6-16. *Deleting the web services you launched earlier*

Now, navigate to the All resources section again from the home portal. Check every item with the resource group type named Container Instance. You should see that there are none. Figure 6-17 shows what this might look like. (We have a container instance here, but it is unrelated.) Since you deleted the services just now, you should not see sklearn-logreg or sklearn-model-dev.

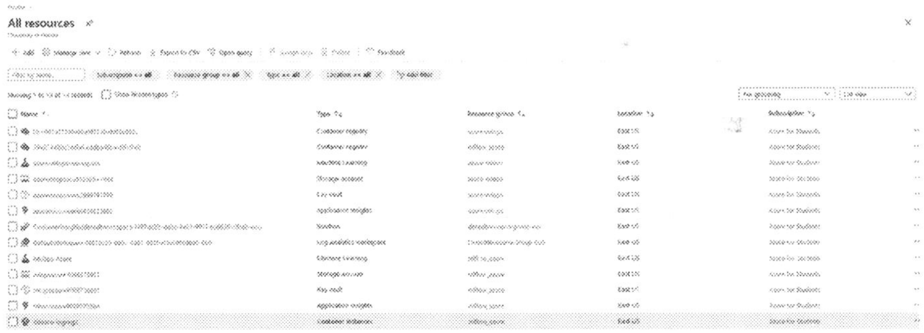

Figure 6-17. *You should not see any resources titled sklearn-logreg or sklearn-model-dev of type container instance. (There is one here, but it is not related to the experiments from above, and only exists to show what a resource with this resource type looks like.)*

If you want to remove services from here, you can simply delete the container instances or other services, as in Figure 6-18.

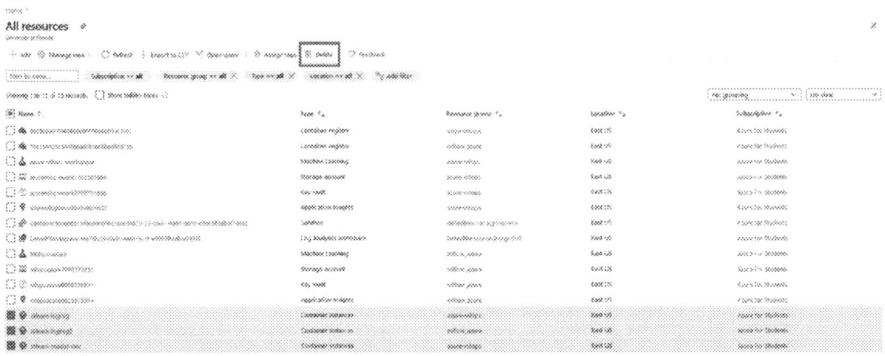

Figure 6-18. *Deleting services manually through the All resources UI*

You can now delete everything else (or just the new resources created for this chapter) in your UI following this same procedure to clean up your Azure workspace.

With that, you now know how to use MLFlow to deploy a model on Microsoft Azure.

It's worth mentioning that Azure has a lot of additional functionality relating to monitoring your machine learning experiments and more, but that might also come with additional costs depending on the depth of functionality you are going after. Be sure to refer to their excellent documentation if you'd like to learn more about Azure and its functionality.

Summary

Like Amazon AWS, Microsoft Azure is a cloud platform that performs many advanced services for a wide range of users. In particular, Azure has a lot of support for operationalizing machine learning models using built-in functionality separate from MLFlow.

In this chapter, you learned how to build a container image for a specific MLFlow model run, deploy it in a development setting/production setting, and query the model on Microsoft Azure.

In the next chapter, we will look at how you can use Google Cloud as a platform to operationalize your MLFlow models. There is no explicit MLFlow support for Google Cloud, so you will be adopting a different approach where you serve the models on a server hosted on Google Cloud and make predictions that way.

CHAPTER 7

Deploying in Google

In this chapter, we will cover how you can use MLFlow and Google Cloud to operationalize your models even without MLFlow providing explicit deployment support for Google Cloud.

More specifically, we will cover how to set up your Google Cloud bucket and virtual machine (used to run the server) and how you can operationalize and query your models.

Introduction

In the previous chapter, we went over how you can deploy your models to Microsoft Azure, manage them through update or delete events, and query them. This time, we will explore how you can operationalize your models using Google Cloud.

MLFlow does not provide explicit support for deploying in Google Cloud like it does with AWS SageMaker and Microsoft Azure, and so you will approach this a bit differently from how you operationalized models in the previous two chapters.

This time, you will use the same model serving functionality that you used in Chapter 4 except you will host it on a Google Cloud machine that is accessible by the Internet. However, deployment is far quicker this way since you don't have to wait for the creation of an endpoint. Furthermore, once you set up the machine, swapping models is very simple, and you can serve multiple models by using different ports.

© Sridhar Alla, Suman Kalyan Adari 2021
S. Alla and S. K. Adari, *Beginning MLOps with MLFlow*,
https://doi.org/10.1007/978-1-4842-6549-9_7

It's worth noting that Google Cloud has an assortment of advanced tools and functionality dedicated to machine learning, such as Kubeflow. Kubeflow is a tool that allows you to essentially integrate your machine learning lifecycles into Kubernetes. And so all your machine learning pipelines are managed through Kubernetes. Kubeflow also integrates into the Google Cloud platform, seeing as how Kubernetes was built by Google. In this chapter, we will just go over how you can deploy MLFlow logged models. We won't get into any of the platform-specific tools that help manage your machine learning lifecycles.

Before you begin, here is an **important prerequisite**:

- Download and install the Google Cloud SDK so you can use the CLI to connect to your server.

In detail, we will go over the following in this chapter:

- **Configuring Google:** This is perhaps the hardest step in this deployment process. First, you set up a bucket and push the contents of your `mlruns` folder to be stored on the cloud.

  Next, you set up the virtual machine that will host your server when you deploy the model. This involves installing Conda and MLFlow.

  Finally, you set up a firewall to allow your server to have inbound access through the default port of 5000 that MLFlow uses so that you can actually connect to this server through your Jupyter notebook.

- **Deploying and querying the model:** Here, you check the IP address, pick a run, and launch the code to serve the model. Then, you query the model and run the batch query script as well.

- **Updating and removing a deployment:** Here, you stop deployment and simply rerun the model serving script with a different model run to fulfill model switching functionality. After you have updated the model, removing the deployment is as easy as stopping the model serving.

- **Cleaning up:** Here, you go through all of the new services you used and delete them all so as not to incur any charges.

With that, let's get started!

Configuring Google

Most of the work that is involved in deploying your models using Google Cloud is actually taken up by the configuration process. Once you set up the storage and the machine to host your model, model serving becomes an extremely easy task. To switch up models, you only need to change up the model run and let MLFlow take care of the rest.

As for where you are storing the models, you will be using Google Cloud Storage to do so. Once again, this fulfills a functionality similar to storing your runs in Amazon S3 buckets or Azure blobs. The purpose of pushing all of your runs to the cloud is so that there is a centralized storage container that holds the models. Now anyone can access them anywhere around the world, and there are no issues with version mismatch where your copy of the run happens to differ with someone else's. In a sense, this is serving the role of a model registry, just without the added functionality of the MLFlow Model Registry.

Bucket Storage

And so, let's begin. First, open up the Google Cloud portal. You should
see something like Figure 7-1. Be aware, though, that Google Cloud is also
constantly being updated, so your portal screen may look different.

Figure 7-1. *What our Google Cloud portal screen looks like*

Notice the scroll bar on the left side of the screen. This is where you can
look at the services Google Cloud provides. Scroll to the section that says
Storage, and click the service named Storage. You should see something
similar to Figure 7-2.

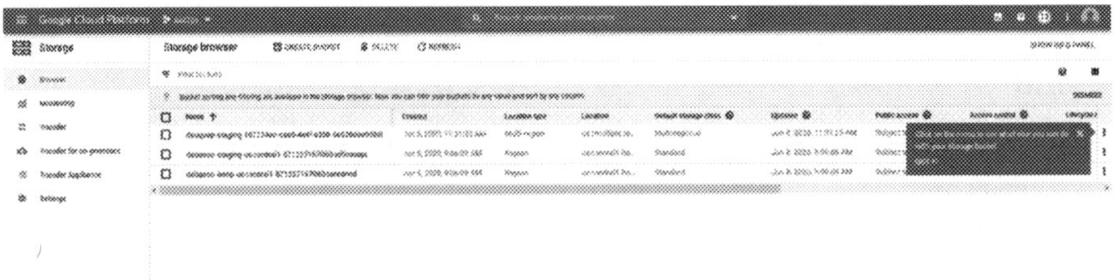

Figure 7-2. *Something similar to what you might see. In your case,
you might not have any buckets here*

Click the button that says CREATE BUCKET. Type in `mlops-storage`.

Next, where it asks for a location type, select the Region option to have the lowest costs. Refer to Figure 7-3.

Figure 7-3. *Specifying the storage option for your bucket. Select Region to keep the costs the lowest, although with the amount of data you are pushing, the actual costs are very little*

Keep the rest of the options as is and click the Create button. You should now see something that looks like Figure 7-4.

Figure 7-4. *What your bucket might look like after creation*

From here, you want to upload your MLFlow experiments (the content of your `mlruns` directory) as folders, so click Upload Folder, and upload all of the folders inside the `mlruns` directory. You can leave out the folder named `.trash`. In our case, we only **uploaded the experiment using scikit-learn** and left the rest out since we won't be using the other experiments.

You should see something like Figure 7-5 when finished.

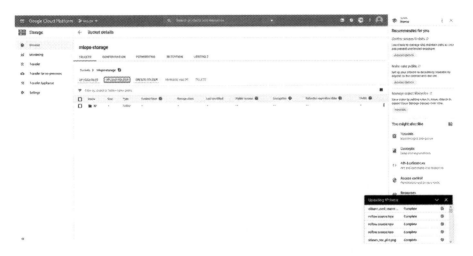

Figure 7-5. *Our bucket after uploading the contents of our mlruns directory. We only uploaded the experiment using scikit-learn to save on costs*

With that, you have finished configuring your storage. The next thing to configure is the virtual machine that will be hosting your model.

Configuring the Virtual Machine

After going back to the portal, scroll to the Compute section and click the Compute Engine option. You should see something like Figure 7-6. You want to make sure you're in the portal for the service titled VM Instances.

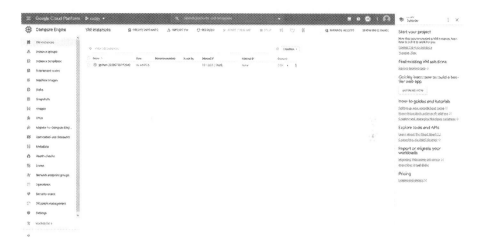

Figure 7-6. *What your VM Instances screen may look like. In our case, we already have another machine running, but that is irrelevant since we are creating a new machine*

Now, click Create Instance and you should see something like Figure 7-7.

Figure 7-7. *The options you can fill in when creating your VM machine instance. You should match the selections shown in the figure to ensure consistency with our results*

In our case, we filled in or selected the options that we want our VM machine to use. We named our machine mlops-server, selected our region (it autoselects a zone for you), and specified that we want to use Ubuntu 18.04 LTS. Finally, at the end, we want to allow HTTPS traffic from the internet.

Finally, when finished, you should be able to see your VM machine on the list of machines. What you want to do now is to open your VM machine instance by clicking the name `mlops-server`. This should take you to a screen that looks like Figure 7-8.

Figure 7-8. *What you should see when you click mlops-server. Notice the box that says SSH. You will use that shortly*

Now look at the box that says SSH. There should be a little down arrow indicating that it is a drop-down list of something. Click that arrow and select the View gcloud command option. Refer to Figure 7-9.

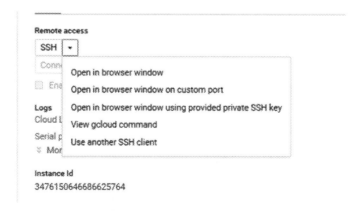

Figure 7-9. *The drop-down options for connecting to this VM instance*

This should take you to a popup window that looks like Figure 7-10. You have two options: running that command in a new instance of the Google Cloud SDK CLI (in our case, we had to search "Google Cloud SDK Shell" and it opened a configured Google Cloud terminal instance), or running it through a shell directly on the portal page itself. You can do either option, as both connect to the VM anyway.

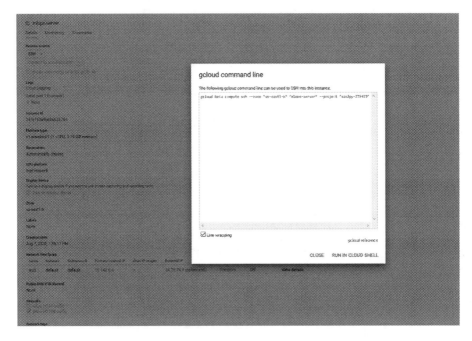

Figure 7-10. *The command that lets you connect to the VM via SSH. You can also run it within the portal page itself if you'd like*

Copy and paste that command in your terminal to connect to the VM. When finished running, you should see something like Figure 7-11, where it opens up a PuTTY instance of the actual shell inside the VM.

Figure 7-11. *The result of running the gcloud command that the portal provided. On the right, you can see a PuTTY terminal where you have the shell open inside the VM*

This is where you must configure your VM so that it can host your MLFlow models.

First, run the following commands:

```
sudo apt update
```

```
sudo apt upgrade
```

Answer "y" to any prompts.

Once finished, you can now install Conda. Without Conda, MLFlow won't be able to reconstruct the environment that the MLFlow model was logged in. This is part of MLFlow's modularization. In the case of SageMaker and Azure, you built containers that, as their name suggests, "contain" these Conda environments already. This way, SageMaker does not have to reinstall any Conda packages once the container is in the cloud. It simply has to run an instance of the container and it already has everything configured.

First, find out how to install Anaconda on Linux by going to its webpage. An install link should be provided. Copy the link and paste it somewhere. You will retrieve that link using a command.

Run the following one at a time:

```
cd /tmp
curl -O https://repo.anaconda.com/archive/
Anaconda3-2020.07-Linux-x86_64.sh
```

You should see something like Figure 7-12.

```
Shumpu@mlops-server:~$ cd /tmp
Shumpu@mlops-server:/tmp$ curl -O https://repo.anaconda.com/archive/Anaconda3-20
20.07-Linux-x86_64.sh
  % Total    % Received % Xferd  Average Speed   Time    Time     Time  Current
                                 Dload  Upload   Total   Spent    Left  Speed
100  550M  100  550M    0     0   165M      0  0:00:03  0:00:03 --:--:--  165M
Shumpu@mlops-server:/tmp$
```

Figure 7-12. *The output of fetching the Anaconda installation script*

Next, let's install Anaconda by running the following. You can type in
bash Anaconda and press Tab to autofill the rest of the script name.

```
bash https://repo.anaconda.com/archive/
Anaconda3-2020.07-Linux-x86_64.sh
```

It should ask you to look through the license agreement. At the end,
answer yes, and press Enter to confirm the default installation location.
Conda should then proceed with the installation. Answer yes to any
further prompts. Once it's done, restart the shell (close the PuTTY client
and rerun the command or cloud shell), and you should now have Conda
fully configured.

As you will now see, Conda has already started the base environment.
Let's create a new environment by running the following code:

```
conda create -n mlflow python=3.7
```

Answer "y" to any following prompts, and you should see something
like Figure 7-13.

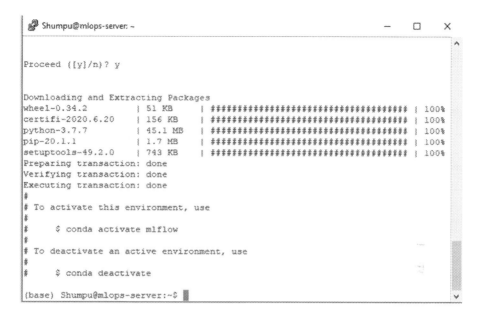

```
 Shumpu@mlops-server: ~                                                      —    □    ×

Proceed ([y]/n)? y

Downloading and Extracting Packages
wheel-0.34.2        | 51 KB    | ################################### | 100%
certifi-2020.6.20   | 156 KB   | ################################### | 100%
python-3.7.7        | 45.1 MB  | ################################### | 100%
pip-20.1.1          | 1.7 MB   | ################################### | 100%
setuptools-49.2.0   | 743 KB   | ################################### | 100%
Preparing transaction: done
Verifying transaction: done
Executing transaction: done
#
# To activate this environment, use
#
#     $ conda activate mlflow
#
# To deactivate an active environment, use
#
#     $ conda deactivate

(base) Shumpu@mlops-server:~$
```

Figure 7-13. *If you see this, then your Conda environment has successfully installed*

Next, you will install the following packages: **mlflow** and **google-cloud-storage**. The former is self-explanatory: you will need MLFlow to do anything with MLFlow. You need google-cloud-storage because you are going to access your runs from the Google storage bucket from earlier.

Run the following:

```
conda activate mlflow
```

```
pip install mlflow google-cloud-storage
```

Running this code should also install all of the dependencies. In the future, should you need to install any more dependencies, it's as simple as activating the mlflow environment and using pip install to get any more packages or update existing packages.

Once it has finished installing everything, you should see something like Figure 7-14.

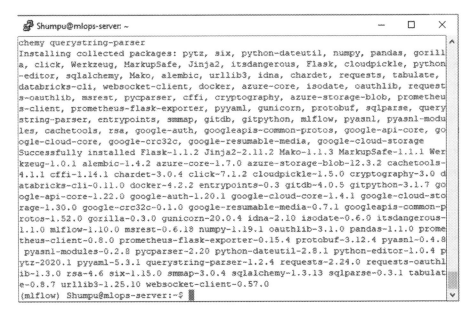

Figure 7-14. *The final output after finishing installing the necessary packages in the Conda environment*

With that, you have fully configured your VM. All that is left is to configure the firewall.

Configuring the Firewall

First, you need to look at the internal IP that your VM instance is using. To do that, run the following:

```
ifconfig
```

You should see something like Figure 7-15.

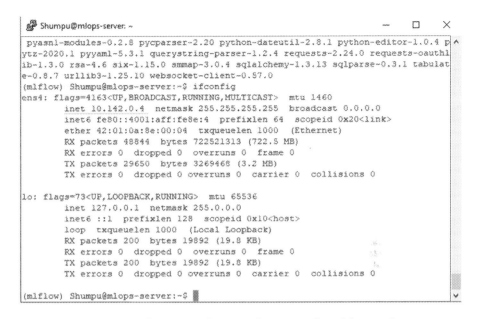

Figure 7-15. *Something similar to what you should see when you run the command. We have highlighted in red where you can find the internal IP of your machine. In our case, it is 10.142.0.4. Yours will be different*

Make a note of the internal IP, which we have highlighted in red. In your case, it will be different.

Now, you must add a firewall to allow access to your server once it is started. Go back to the portal, scroll to the section that says Networking, and click the VPC Networks option. You should see something like Figure 7-16.

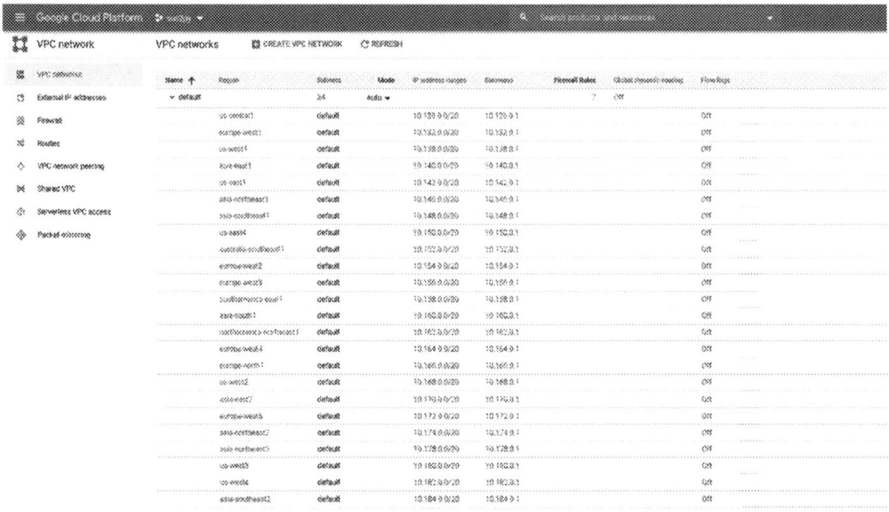

Figure 7-16. *The VPC Networks module in the portal. Click the Firewall option to look at the firewall options*

Now, click Firewall and then click Create Firewall Rule. Namely, you want to enter the following values:

- **Name**: mlflow-server

- **Target tags**: mlops-server, http-server, https-server

- **Source IP ranges**: 0.0.0.0/0

- **Protocols and Ports**: Check TCP and type 5000

If you made a mistake, you can edit the firewall rules. You should see something like Figure 7-17.

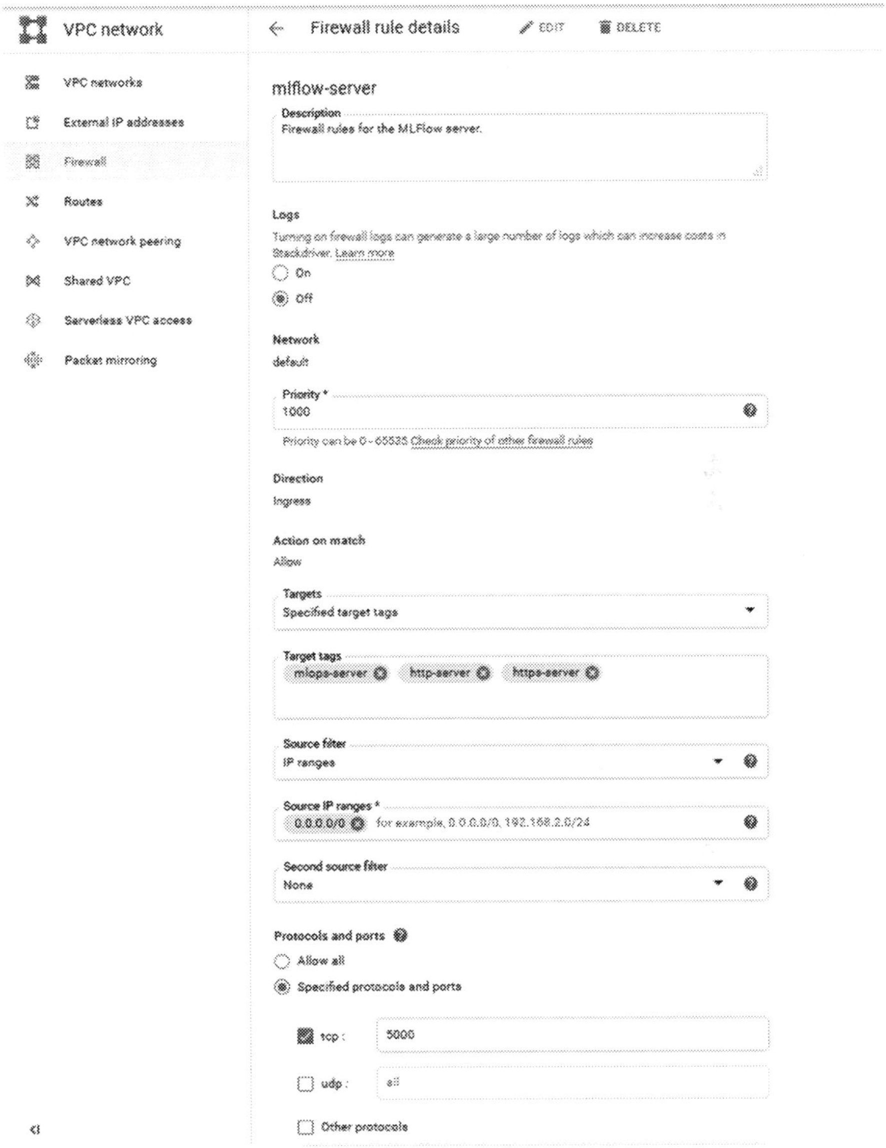

Figure 7-17. *What your firewall configuration should look like. We have autofilled the values with our own*

Now click Create. You are done configuring the firewall and configuring everything else in Google Cloud. Now you can move on to deploying your model.

Deploying and Querying the Model

With your virtual machine fully configured, it's time to deploy your model.

Make sure you still have that internal IP logged in. Go back to the PuTTY client and now enter the following command:

```
mlflow models serve -m gs://mlops-storage/EXPERIMENT_ID/RUN_ID/
artifacts/MODEL_NAME -h 10.142.0.4
```

Our command looks like the following. We simply took the first run in the Google Storage bucket.

```
mlflow models serve -m gs://mlops-storage/8/1eb809b446d949d5a70
a1e22e4b4f428/artifacts/log_reg_model -h 10.142.0.4
```

You should see something like Figure 7-18.

```
Shumpu@mlops-server: ~                                            —    □    ×

0.15.4 protobuf~3.12.4 pycparser-2.20 python-dateutil~2.8.1 python-editor~1.0.4 ^
pytz-2020.1 pyyaml-5.3.1 querystring-parser-1.2.4 requests-2.24.0 requests-oauth
lib-1.3.0 six-1.15.0 smmap-3.0.4 sqlalchemy-1.3.13 sqlparse-0.3.1 tabulate-0.8.7
 urllib3-1.25.10 websocket-client-0.57.0


#
# To activate this environment, use
#
#     $ conda activate mlflow-dd9a9304d9fe43d69fd90c7ad7a9f26ba0cae2aa
#
# To deactivate an active environment, use
#
#     $ conda deactivate

2020/08/07 18:30:34 INFO mlflow.pyfunc.backend: === Running command 'source /hom
e/Shumpu/anaconda3/bin/../etc/profile.d/conda.sh && conda activate mlflow-dd9a93
04d9fe43d69fd90c7ad7a9f26ba0cae2aa 1>&2 && gunicorn --timeout=60 -b 10.142.0.4:5
000 -w 1 ${GUNICORN_CMD_ARGS} -- mlflow.pyfunc.scoring_server.wsgi:app'
[2020-08-07 18:30:34 +0000] [26836] [INFO] Starting gunicorn 20.0.4
[2020-08-07 18:30:34 +0000] [26836] [INFO] Listening at: http://10.142.0.4:5000
(26836)
[2020-08-07 18:30:34 +0000] [26836] [INFO] Using worker: sync
[2020-08-07 18:30:34 +0000] [26846] [INFO] Booting worker with pid: 26846       v
```

Figure 7-18. *This is what your output should look like if it successfully built the Conda environment and is now serving the model*

There's only one more step that remains before you can successfully make predictions with this model. You must now see what your external IP is. To do so, go back to the VM Instances page to find your VM machine. You should see something like Figure 7-19.

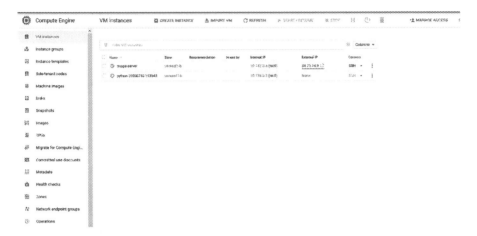

Figure 7-19. *The VM Instances section in the portal should display the external IP of your server. We have highlighted ours in red, but yours is most likely something different*

Once you have the external IP address, copy it down somewhere.

Now you can start up your Jupyter notebook and query this model.

In a Jupyter notebook cell, run the following. Make sure you have the data folder in the same directory as this notebook, and that the data folder contains the creditcard.csv file:

```
import pandas as pd

import seaborn as sns

import matplotlib.pyplot as plt

from sklearn.preprocessing import StandardScaler
from sklearn.model_selection import train_test_split
from sklearn.metrics import roc_auc_score, accuracy_score,
confusion_matrix
```

```python
import numpy as np

import subprocess
import json

df = pd.read_csv("data/creditcard.csv")
```

Next, you define your query() function that you will use to get model predictions:

```python
def query(input_json):
    proc = subprocess.run(["curl",  "-X", "POST", "-H",
"Content-Type:application/json; format=pandas-split",
                      "--data", input_json,
                      "http://34.75.74.9:5000/invocations"],
                      stdout=subprocess.PIPE, encoding='utf-8')

    output = proc.stdout
    preds = json.loads(output)
    return preds
```

Notice that the IP is now http://34.75.74.9:5000/invocations. Basically, your IP should take the form of http://YOUR_EXTERNAL_ IP:5000/invocations, replacing the placeholder with the external IP address of your VM.

Let's now query your model:

```python
input_json = df.iloc[:80].drop(["Time", "Class"],
axis=1).to_json(orient="split")
pd.DataFrame(query(input_json)).T
```

Altogether, you should see something like Figure 7-20.

```
In [1]:    1  import pandas as pd
           2  import mlflow
           3  import mlflow.sklearn
           4
           5  import seaborn as sns
           6
           7  import matplotlib.pyplot as plt
           8
           9  from sklearn.preprocessing import StandardScaler
          10  from sklearn.model_selection import train_test_split
          11  from sklearn.metrics import roc_auc_score, accuracy_score, confusion_matrix
          12
          13  import numpy as np
          14
          15  import subprocess
          16  import json
          17
          18  df = pd.read_csv("data/creditcard.csv")

In [15]:   1  def query(input_json):
           2      proc = subprocess.run(["curl", "-X", "POST", "-H", "Content-Type:application/json; format=pandas-split",
           3                             "--data", input_json, "http://54.75.74.9:5000/invocations"],
           4                             stdout=subprocess.PIPE, encoding='utf-8')
           5
           6      output = proc.stdout
           7      preds = json.loads(output)
           8      return preds

In [17]:   1  input_json = df.iloc[:80].drop(["Time", "Class"], axis=1).to_json(orient="split")
           2  pd.DataFrame(query(input_json)).T
Out[17]:
```

```
   0  1  2  3  4  5  6  7  8  9  ...  70  71  72  73  74  75  76  77  78  79
0  1  0  1  1  1  0  0  1  1  0  ...   1   0   0   0   0   0   1   1   0   0
```

1 rows × 80 columns

Figure 7-20. *The output of querying the model with the first 80 rows of your data frame*

As expected, the predictions aren't correct because you did not scale the data before querying the model with it. However, you have verified that you have queried the correct address and that the model is able to return predictions.

Now run the following cells:

```
normal = df[df.Class == 0].sample(frac=0.5, random_state=2020).
reset_index(drop=True)
anomaly = df[df.Class == 1]

normal_train, normal_test = train_test_split(normal, test_size
= 0.2, random_state = 2020)
anomaly_train, anomaly_test = train_test_split(anomaly,
test_size = 0.2, random_state = 2020)
```

```
scaler = StandardScaler()
scaler.fit(pd.concat((normal, anomaly)).drop(["Time", "Class"],
axis=1))

test = pd.concat((normal.iloc[:1900], anomaly.iloc[:100]))
true = test.Class
test = scaler.transform(test.drop(["Time", "Class"], axis=1))
preds = []

batch_size = 80
for f in range(25):
    print(f"Batch {f}", end=" - ")

    sample = pd.DataFrame(test[f*batch_size:(f+1)*batch_size]).
    to_json(orient="split")

    output = query(sample)
    resp = pd.DataFrame([output])
    preds = np.concatenate((preds, resp.values[0]))

    print("Completed")

eval_acc = accuracy_score(true, preds)
eval_auc = roc_auc_score(true, preds)

print("Eval Acc", eval_acc)
print("Eval AUC", eval_auc)
```

Once finished, you should see something like Figure 7-21.

```
In [19]:  1  test = pd.concat((normal.iloc[:1900], anomaly.iloc[:100]))
          2  true = test.Class
          3  test = scaler.transform(test.drop(["Time", "Class"], axis=1))
          4  preds = []
          5
          6  batch_size = 80
          7  for f in range(25):
          8      print(f"Batch {f}", end=" - ")
          9
         10      sample = pd.DataFrame(test[f*batch_size:(f+1)*batch_size]).to_json(orient="split")
         11
         12      output = query(sample)
         13      resp = pd.DataFrame([output])
         14      preds = np.concatenate((preds, resp.values[0]))
         15
         16      print("Completed")
         17
         18  eval_acc = accuracy_score(true, preds)
         19  eval_auc = roc_auc_score(true, preds)
         20
         21  print("Eval Acc", eval_acc)
         22  print("Eval AUC", eval_auc)

Batch 0 - Completed
Batch 1 - Completed
Batch 2 - Completed
Batch 3 - Completed
Batch 4 - Completed
Batch 5 - Completed
Batch 6 - Completed
Batch 7 - Completed
Batch 8 - Completed
Batch 9 - Completed
Batch 10 - Completed
Batch 11 - Completed
Batch 12 - Completed
Batch 13 - Completed
Batch 14 - Completed
Batch 15 - Completed
Batch 16 - Completed
Batch 17 - Completed
Batch 18 - Completed
Batch 19 - Completed
Batch 20 - Completed
Batch 21 - Completed
Batch 22 - Completed
Batch 23 - Completed
Batch 24 - Completed
Eval Acc 0.9915
Eval AUC 0.915
```

Figure 7-21. *The results of running your batch query script*

Updating and Removing a Deployment

Updating the model deployment is extremely easy. With how you set it up, it's only a matter of quitting the model serving command (Ctrl-C), and rerunning the command with a different run ID.

Let's try deploying a different run. In your case, check your Google Storage bucket and pick the second run.

In our case, we ran the following:

```
mlflow models serve -m gs://mlops-storage/8/3862eb3bd89b43e8ace
610c521d974e6/artifacts/log_reg_model -h 10.142.0.4
```

As you can see in Figure 7-22, it successfully deployed, and we can simply query it using the same script.

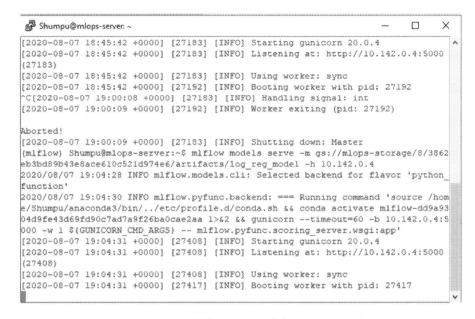

Figure 7-22. Deploying a different model run using the same command convention

As for removing a deployment, all you have to do is just cancel the command with Ctrl-C and your deployment is now cancelled.

With that, you now know how to serve models, switch a model and deploy a different one, and remove a deployment by simply canceling the model serving command.

Cleaning Up

It's time to delete every instance of a service that you created so that you won't incur any charges. Here's a list of all of the services you used:

- Google Cloud Storage Bucket

- Compute Engine VM Instance

- Networking Firewall Rule

Beginning with your VM Instance, you want to click STOP to first stop the VM from running. You should see something like Figure 7-23 depending on where you access this VM.

Figure 7-23. *The VM instance after stopping it*

After that, you can simply click DELETE to remove the VM. Stopping the VM only ensures that you won't be billed for CPU/GPU utilization, but it won't stop any charges that result from services linked to the VM.

Next, let's go to the Storage bucket. Simply check your bucket and click DELETE to remove this storage. Refer to Figure 7-24.

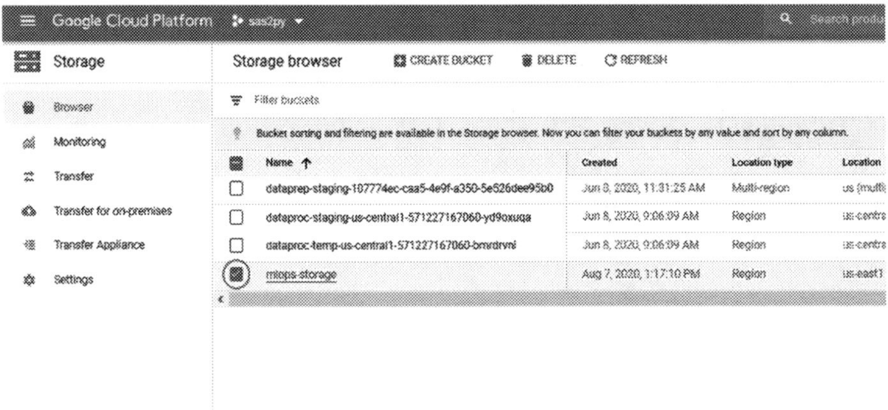

Figure 7-24. *Removing your storage bucket*

Lastly, you may remove the firewall rule as well, but be sure to not remove any other rules that you might have in there.

With that, your workspace should be cleaned up, and there shouldn't be any more services that may incur charges.

Summary

Google Cloud is a cloud platform that provides many advanced services for a wide range of users. While MLFlow does not explicitly provide support for deployment for Google Cloud, you are still able to operationalize your models using MLFlow's model serving functionality and Google Cloud's compute engine to serve the models on the cloud.

In this chapter, you learned how to set up Google Cloud so that it can deploy your models on a virtual machine. In particular, you looked at how you can push your MLFlow runs to a bucket, how you can set up the Conda environment on a virtual machine, how you can set up a firewall to allow your model to be accessed in order to be queried, and how you can manage your deployments by simply switching out run IDs (and experiment IDs where appropriate).

In the Appendix, you can look at how Databricks helps you operationalize your models and manage them through the use of a model registry.

APPENDIX

Databricks

In this appendix, we will cover what Databricks is as well as how you can utilize its built-in MLFlow functionality to log MLFlow runs within Databricks itself, how to deploy models from Databricks to Azure, and how the MLFlow model registry works in Databricks.

Introduction

Databricks is an open platform and cloud service that provides interoperability with other popular AI and data services like AWS and Microsoft Azure. Databricks also created Apache Spark, Delta Lake, and MLFlow (see Chapter 4 to learn what MLFlow is).

Before we begin, you will need a Databricks account. You have the option of creating a "community edition" account, which is free to users but is limited in its functionality. You will be able to use basic MLFlow functionality on top of whatever Python functionality you have (PySpark is supported, for example), but you will not be able to use the model registry functionality.

To sign up for one, head on over to this website: `https://community.cloud.databricks.com/`.

Otherwise, you will have to pay to be able to use Databricks by choosing a subscription plan for your account.

© Sridhar Alla, Suman Kalyan Adari 2021
S. Alla and S. K. Adari, *Beginning MLOps with MLFlow*,
https://doi.org/10.1007/978-1-4842-6549-9

With Databricks, you can integrate with Amazon AWS or Microsoft Azure. If you choose to subscribe to a plan from Databricks, you will be integrating with AWS. However, you can also deploy Databricks in Azure, which you can find more information about here:
`https://azure.microsoft.com/en-us/services/databricks/`.

Be warned, although Microsoft Azure does offer a free, 14-day trial of Databricks, you cannot create clusters without upgrading to the premium version of Azure Databricks (with a paid Azure subscription).

In this appendix, we will be using the **community edition** of Databricks, which is free to sign up for an use. The only exception here is the section in which we cover the model registry, which seems to only be available to premium Databricks users.

In detail, we will go over the following:

- **Logging MLFlow runs within Databricks:** You can run your Jupyter notebooks within Databricks itself, which provides functionality to import your old notebooks. For this part, you will import your notebook from Chapter 4 where you conduct experiments using scikit-learn. All runs will be logged within Databricks.

- **MLFlow UI:** Databricks has a built-in MLFlow UI that allows you to see all of your runs per experiment just as you would in the browser. You will look at your experiment using this UI and inspect a run that you will log.

- **Deploying to AWS/Azure:** Depending on what you integrate with, you can deploy your models to one of these services. In this chapter, we will be deploying to Microsoft Azure.

- **MLFlow Model Registry:** With premium Databricks (non-community edition), you have the added capability of having a model registry. Here, we will go over what the model registry is and how it works.

With that, let's get started!

Running Experiments in Databricks

Once you have Databricks set up, whether in community edition or otherwise, you should be greeted with a home screen that looks somewhat like Figure A-1.

Figure A-1. *The Databricks home screen. If you have the community edition, you won't have the Models tab on the navigation bar to the left, but otherwise it should look about the same*

Where it says Common Tasks, go down until you see the option titled New MLFlow Experiment. Click this option.

You can type in any other name you like, but you should see something like Figure A-2.

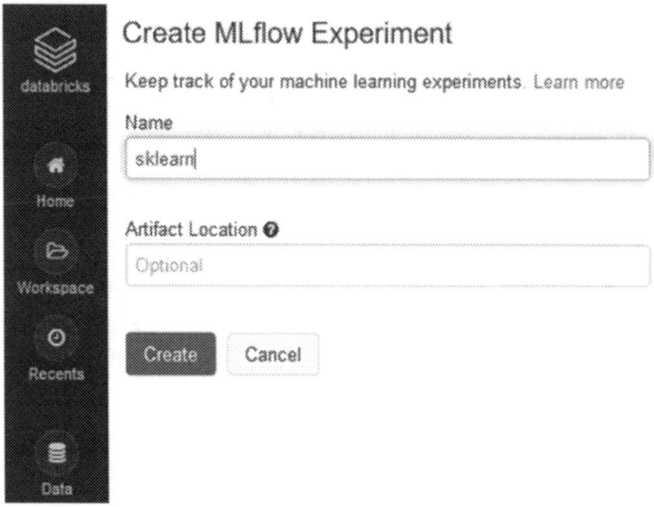

Create MLflow Experiment

Keep track of your machine learning experiments. Learn more

Name

sklearn|

Artifact Location ❓

Optional

[Create] [Cancel]

Figure A-2. *The screen you should see when creating an MLFlow experiment*

Go ahead and click Create. You should now see the MLFlow UI displaying the details of this experiment. Of course, there are no runs since you just created it. You should see something like Figure A-3.

Figure A-3. *The screen displayed after experiment creation. Note that the experiment name is now /Users/sadari@bluewhale.one/sklearn. Be sure to make note of this as this is the full experiment name you will use when setting the experiment in the code*

Something important to mention is that the experiment name in this case is not sklearn, but rather it is /Users/sadari@bluewhale.one/sklearn *in its entirety*. Whatever you see is what you will be using when setting the experiment in the notebook code.

With that, simply click Databricks to return to the home screen.

You now have two choices:

1. Create a new notebook and fill in the cells from scratch.

2. Import your MLFlow scikit-learn notebook from Chapter 4.

In this chapter, you will be importing the MLFlow scikit-learn notebook, but you will be making a few changes in order to ensure that it is adapted to work with Databricks.

Before you even begin with the notebook, however, you need to create the **cluster** that will run your notebook code. To do this, click the New Cluster option, and you should see something like Figure A-4.

Figure A-4. *Cluster creation UI in the community edition of Databricks. Here, the name and the 7.2 ML runtime are autofilled*

Make sure that you have the same runtime as in Figure A-4, or at least something that has "ML" in the runtime name. Once finished, click the Create Cluster option.

After that, you'll be taken to a UI that shows all the clusters you have. Refresh if the cluster does not immediately show up. This can take a bit, so in the meantime, let's head back to the home screen.

At this point, you can proceed with your notebook. On the left navigation pane, click Home > Users (if it's not selected for you), and then click your username to open a dropdown window. You should see something like Figure A-5.

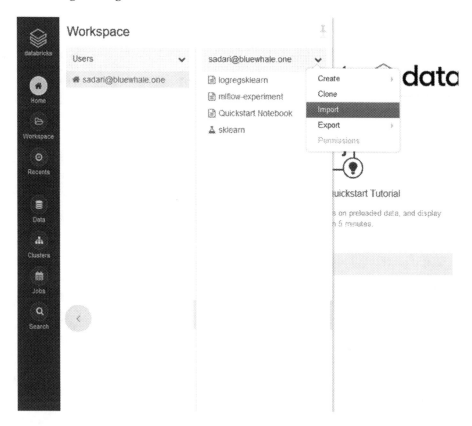

Figure A-5. *Home menu that allows you to import a notebook. Don't worry about the other files you see here; you are likely to only have the experiment named sklearn and perhaps the Quickstart Notebook file*

Click Import and navigate to your MLFlow notebook from Chapter 4 (if you have one just for scikit-learn, that is preferable).

You will now be taken to a notebook with all the contents of the notebook you just imported, except for the outputs.

Before you get to run this, you must import your data. To do this, refer to Figure A-6. You must click File ➤ Upload Data in the dropdown menu.

Figure A-6. *Uploading the data so that it can be accessed by this notebook*

Leaving everything else as is, click Browse and locate and upload your credit card dataset (`creditcard.csv`).

This will take some time to upload due to the size of the file, but once it is all done, click Next, which will give you code samples that tell you how to import this file. Make sure you have selected pandas. You can now paste this code and try to run it. In our case, we had an error stating that the file did not exist, so we instead loaded it with Spark and converted it into a pandas data frame, which does work for some reason given the same file path.

Before you can execute anything, make sure that the cluster has finished building. Above the first cell in the notebook, you'll notice a bar that says "detached." Click it and you should see your cluster available here. If the cluster is ready to use, there should be a green dot beside it. Otherwise, it will have the loading circle indicating that it's still configuring.

Go ahead and click the cluster. Once it is finished, you should see something like Figure A-7.

Figure A-7. *An indication that the cluster is ready to use. If you see the green dot, you can now execute the cells in the notebook*

Now you can begin with the modifications to the code. Let's start with the import statements. Change the first cell to look like the following:

```
import numpy as np
import pandas as pd
import matplotlib #
import matplotlib.pyplot as plt
import seaborn as sns

import sklearn #
from sklearn.linear_model import LogisticRegression
from sklearn.model_selection import train_test_split
from sklearn.preprocessing import StandardScaler
```

```
from sklearn.metrics import roc_auc_score, plot_roc_curve,
confusion_matrix, accuracy_score
from sklearn.model_selection import KFold

import pyspark
from pyspark.sql import SparkSession
from pyspark import SparkConf, SparkContext
import os

import mlflow
import mlflow.sklearn

print("Numpy: {}".format(np.__version__))
print("Pandas: {}".format(pd.__version__))
print("matplotlib: {}".format(matplotlib.__version__))
print("seaborn: {}".format(sns.__version__))
print("Scikit-Learn: {}".format(sklearn.__version__))
print("MLFlow: {}".format(mlflow.__version__))
print("PySpark: {}".format(pyspark.__version__))
```

Here, you have added extra import statements so that you import PySpark.

Create a new cell beneath your first cell, adding the following:

```
os.environ["SPARK_LOCAL_IP"]='127.0.0.1'
spark = SparkSession.builder.master("local[*]").getOrCreate()
spark.sparkContext._conf.getAll()
```

You should see something like Figure A-8 when executed.

Figure A-8. *Running the first two cells and ensuring you have a Spark context*

The next cell should be where you were loading the pandas data frame. Change it to be just the following:

```
df = spark.read.csv("/FileStore/tables/creditcard.csv",
header = True, inferSchema = True).toPandas()
df = df.drop("Time", axis=1)
```

If you run this cell and the next, which should be df.head(), you should see something like Figure A-9.

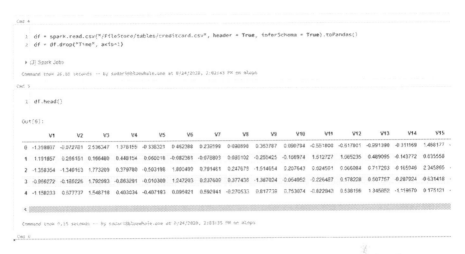

Figure A-9. *Ensuring that you have successfully loaded the data frame in PySpark and have converted it to pandas*

At this point, simply run the rest of the code up until the cell where you actually start the MLFlow run.

You must split up this cell to ensure everything logs to the same run. And so, you can create a new cell if you wish, with the following content:

```
sk_model = LogisticRegression(random_state=None, max_iter=400,
solver='newton-cg')
```

```
mlflow.set_experiment("/Users/sadari@bluewhale.one/sklearn")
```

```
train(sk_model, x_train, y_train)
```

Here are the next three cells. Each text box is supposed to be its own cell:

```
evaluate(sk_model, x_test, y_test)
```

```
mlflow.sklearn.log_model(sk_model, "log_reg_model")
```

```
mlflow.end_run()
```

Together, they should look like Figure A-10.

```
Cmd 15

1  sk_model = LogisticRegression(random_state=None, max_iter=400, solver='newton-cg')
2
3  mlflow.set_experiment("/Users/sadari@bluewhale.one/sklearn")
4
5  train(sk_model, x_train, y_train)


Cmd 16

1  evaluate(sk_model, x_test, y_test)


Cmd 17

1  mlflow.sklearn.log_model(sk_model, "log_reg_model")


Cmd 18

1  mlflow.end_run()
```

Figure A-10. *Splitting up the code to log the relevant metrics and artifacts to ensure everything ends up in the same run. It seems counterintuitive, but lumping it all under the same run with mlflow. start_run() seems to cause the runs to fail*

Now, run these cells. You should now see all of this logged in the experiment.

To view your runs, click Workspace in the navigation pane, and then click sklearn and the experiment name. You should see a run logged there. Click it, and you should see something like Figure A-11, with all the metrics and artifacts logged successfully.

Figure A-11. *Viewing the metrics and artifacts of the run and ensuring they were logged successfully*

With that, you are now ready to deploy. Logging MLFlow runs is as simple as in Databricks. One of the added benefits of Databricks is that it integrates Spark within its functionality, so if you primarily want to log PySpark models, Databricks might be ideal for you.

Deploying to Azure

Since we have already looked at how to deploy to Azure, we will get straight to the point. If you would like to explore this process in more detail, refer to Chapter 6.

Connecting to the Workspace

In this step, you are simply connecting to an existing workspace through Databricks. It's important to note that Databricks does not have azureml-sdk installed, so you must do so yourself. Luckily, Jupyter allows you to do this in a cell, so simply run the following:

```
!pip install azureml-sdk
```

Next, run the following, replacing all the placeholders with your own corresponding values:

```
import azureml
from azureml.core import Workspace

workspace_name = "databricks-deploy" # Your workspace name
workspace_location="East US" # Your region
resource_group = "azure-mlops" #Your resource group
subscription_id = "xxxxxxxx-xxxx-xxxx-xxxx-xxxxxxxxxxxx"
# Your subscription ID above

workspace = Workspace.create(name = workspace_name,
                             location = workspace_location,
                             resource_group = resource_group,
                             subscription_id = subscription_id,
                             exist_ok=True)
```

When you run this, you should see something like Figure A-12 asking you for authentication. Simply follow the instructions and you should be good to go.

```
     Cmd 28
      1  import azureml
      2  from azureml.core import Workspace
      3
      4  workspace_name = "databricks-deploy"
      5  workspace_location="East US"
      6  resource_group = "azure-mlops"
      7  subscription_id = ████████████████████
      8
      9  workspace = Workspace.create(name = workspace_name,
     10                               location = workspace_location,
     11                               resource_group = resource_group,
     12                               subscription_id = subscription_id,
     13                               exist_ok=True)

     Performing interactive authentication. Please follow the instructions on the terminal.
     To sign in, use a web browser to open the page https://microsoft.com/devicelogin and enter the code F46ABGJK7 to authenticate.
```

Figure A-12. *The cell asking for authentication as you attempt to connect to an existing workspace. Follow the instructions, and the cell should finish with the statement, "Deployed Workspace with name databricks-deploy. Took __ seconds"*

Once this finishes, you can proceed with building and pushing a container image using MLFlow functionality. Before you do that, make sure to keep track of your run ID (you should be able to see this in Figure A-11), and copy that information in the cell below:

```
run_id = "dabea5a03050455aa5ad4a61fa548093"
model_name = "log_reg_model"
model_uri = f"runs:/{run_id}/{model_name}"
```

Next up are the two cells with MLFlow code to build and push the container image:

```
import mlflow.azureml

model_image, azure_model = mlflow.azureml.build_image
                      (model_uri=model_uri, workspace=workspace,
                           model_name="sklearn_logreg",
                           image_name="model",
                           description="SkLearn LogReg
                           Model for Anomaly Detection",
                           synchronous=False)

model_image.wait_for_creation(show_output=True)
```

Together, the cells should look like Figure A-13.

Figure A-13. *The three cells from above and their outputs. Here, you specify a model run and then build and push a container to Azure based on that model*

With this step finished, you are ready to deploy the model using MLFlow Azure.

To do so, simply run the following:

```
azure_service, azure_model = mlflow.azureml.deploy(model_uri,
                            workspace,
                            service_name="sklearn-logreg",
                            model_name="log-reg-model",
                            synchronous=True)
```

With that, let's now check the URI that you will use to query, just to ensure that it has successfully deployed:

```
azure_service.scoring_uri
```

Upon success, you should see something that looks like Figure A-14 for both output cells.

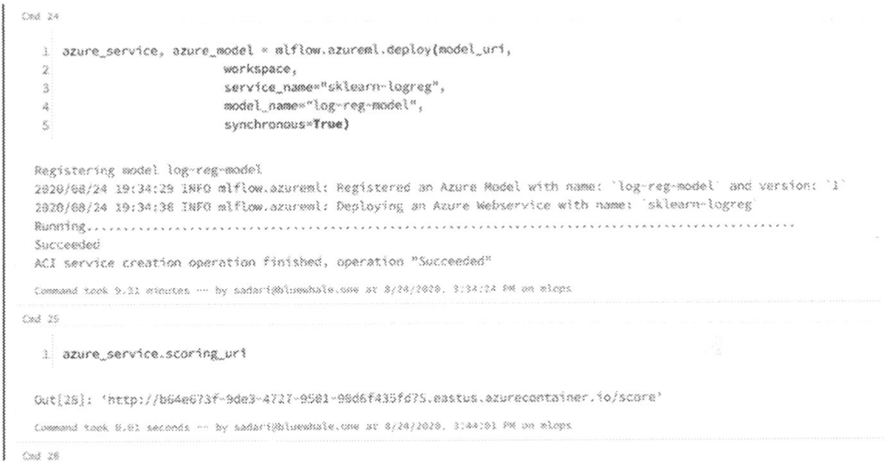

Figure A-14. *The output of deploying the model as well as checking the scoring URI of the service*

Since there is a URI, you know that your model's been deployed successfully. You can move on to the querying process now.

Querying the Model

Before you make any predictions with your model, you need to define a query function:

```
import requests
import json

def query(scoring_uri, inputs):

    headers = {
    "Content-Type": "application/json",
    }
```

```
    response = requests.post(scoring_uri, data=inputs,
    headers=headers)
    preds = json.loads(response.text)
    return preds
```

Let's use your batch query code to query your deployed model and get some relevant metrics. Fortunately, you should already have your scaler object from earlier when you processed the data in the MLFlow experiment.

Simply run the following:

```
test = pd.concat((normal.iloc[:1900], anomaly.iloc[:100]))
true = test.Class
test = scaler.transform(test.drop(["Class"], axis=1))
preds = []

batch_size = 80
for f in range(25):
    print(f"Batch {f}", end=" - ")

    sample = pd.DataFrame(test[f*batch_size:(f+1)*batch_size]).
    to_json(orient="split")

    output = query(scoring_uri=azure_service.scoring_uri,
    inputs=sample)
    resp = pd.DataFrame([output])
    preds = np.concatenate((preds, resp.values[0]))

    print("Completed")

eval_acc = accuracy_score(true, preds)
eval_auc = roc_auc_score(true, preds)

print("Eval Acc", eval_acc)
print("Eval AUC", eval_auc)
```

Your output should look somewhat like Figure A-15.

Figure A-15. *The output of the batch query script. If you cannot see an output past batch 20, resize the output by holding the little arrow on the bottom right*

With that, you now know how to log MLFlow runs in Databricks and deploy models to a cloud platform.

To delete the deployment, simply run the following:

```
azure_service.delete()
```

Be sure to delete all the resources that you created for this deployment as well.

The procedure for AWS is very similar to what you did in Chapter 5, but you just need to set up AWS to allow Databricks to access it.

Databricks has tutorials on how you can accomplish all of that as well. One of the perks of Databricks is that they have extensive documentation about almost everything, especially MLFlow.

MLFlow Model Registry

In this section, we will briefly discuss the model registry. To use the model registry, you do need a premium subscription to Databricks and whatever cloud platform service you choose to deploy Databricks on (either AWS or Azure).

With MLFlow, Databricks provides built-in model registry functionality so that users can define what stage a particular model is in. The MLFlow Model Registry allows for greater collaboration between various teams, letting them all develop and maintain models at various stages in the model life cycle and manage them all in a centralized, organized region.

The user is in control of the lifecycle stage changes (experimentation, testing, production) of the models with options between automatic and manual control. The MLFlow Model Registry tracks the history of the model and allows for governance in managing who is able to approve changes.

Some concepts to know:

- **Registered model:** Once registered in the MLFlow Model Registry, it has a unique name, version, stage, and more.

- **Stage:** Some preset stages are None, Staging, Production, and Archived. The user can also create custom stages for each model version to represent its lifecycle. Model stage transitions are either *requested* or *approved*, depending on the user's level of management.

- **Description:** The user can annotate the model for the team.

- **Activities:** MLFlow records a registered model's activities, providing a history of the model's stages.

Some features include

- **Central repository:** Register MLFlow models to a centralized location.

- **Model versioning:** Keep track of the version history of models. Now, a model built for a specific task can have several versions.

- **Model stage:** Model versions have stages to represent the cycle as a whole. Together with model versioning, older model versions can gradually become phased out while the newest versions are sent to staging first, for example.

- **Model stage transitions:** Respond to new changes and events with automation. Training scripts can be automated to train new models automatically and assign them to staging, for example.

- **CI/CD workflow integration:** Monitor changes to the CI/CD pipelines as new versions are registered and have their deployment stages changed. This allows for better governance over the deployment process.

- **Model serving:** MLFlow models can be served on Databricks through REST APIs, on top of deploying them on a cloud service like AWS or Azure.

With that, let's look at how you can register your model in Databricks. First, head over to your MLFlow experiment and pick a run. Scroll down to artifacts and click the folder that contains your model. **If you don't have premium Databricks, you won't be able to see this Register Model button**. If you click the button and click Create New Model in the dropdown menu, you will see something like Figure A-16.

Figure A-16. Registering a MLFlow model

Once finished, the Register Model button should be replaced by a green checkmark and a link to the model version page of this specific model.

On this page, you can set the model's stage, which is one of None, Staging, Production, or Archived if you're only using preset stages. Furthermore, you can add a description to this specific model.

On top of that, you can also request to change the model's stage (and add an optional comment to add some context), which can be approved, rejected, or canceled.

This allows you to now keep better track of your models by knowing their present stages. There is also support for model versioning, so there can be multiple versions of the model, with the possibility of setting a model stage for each, which you can view at once.

To view all the models that you registered, you can simply click the Models tab in Databricks, as shown in Figure A-17.

Figure A-17. *The navigation pane on the left side of premium Databricks, deployed in Azure in this instance, with the Models tab that will take you to the model registry*

With the model registry that you looked at in prior chapters, where it's just putting the models in a centralized area, you don't have this type of functionality. If you were to implement this, it would have to be through an external program, although it's actually a relatively simple task considering how everything is modularized for you.

With regular MLFlow, this requires you to have a MLFlow server that saves the runs in a mysql, myssl, sqlite, or postgresql dialect. Then, when you open the UI that pertains to this specific server's storage, you can register models and have all of the MLFLow Model Registry functionality.

All of that can get pretty complicated, so Databricks takes care of it all for you, if you have the premium version of Databricks and a subscription to either AWS or Azure, whichever platform you deployed Databricks to.

And that's all there is to the MLFlow Model Registry in Databricks.

With that, you now know how to run Jupyter notebooks in Databricks, how to log MLFlow runs and conduct experiments, and how to deploy your models to a cloud platform.

Summary

Databricks is a cloud platform that integrates with Amazon AWS or Microsoft Azure. As the creator of MLFlow, Databricks integrates MLFlow functionality into its services, allowing you to run all the MLFlow experiments you'd like to on the cloud. Furthermore, it also takes care of the mechanisms behind running a model registry for you, allowing you to take full advantage of MLFlow on the cloud.

In this appendix, you learned how to import your existing notebook, create a MLFlow experiment, and log your own MLFlow runs. On top of that, you also looked at deploying this model to Azure within Databricks itself, and you looked at the model registry and how it works in Databricks.

With this, you now know how to take your existing machine learning experiments and operationalize them very easily with MLFlow. Furthermore, you also know how to deploy your models to three different cloud platforms: Amazon AWS, Microsoft Azure, and Google Cloud. With this chapter, you've also added Databricks to that list, although it's mostly for running your MLFlow experiments on. That being said, you can definitely run MLFlow experiments and log your runs on the other cloud platforms; it's just far easier to do so within Databricks.

Index

A, B

Analyze data, 80
AWS SageMaker
    Amazon ECR, 237
    attributes, 235
    configuration, 232–234
    container, 237
    deploy a model, 238–243
    ECR repository list, 238
    mlruns directory, 235, 236
    predictions, 243–245, 247
    removing deployed
        model, 250, 251
    switching models, 247–249

C

Command line interface (CLI), 229
Continuous delivery, 87
Continuous integration/
    continuous delivery of
    pipelines, 88
    automated model building, 106
    automated training
        pipeline, 111
    data analysis, 106
    deploy pipeline, 110
    feature store, 105

model registry, 111
model services, 111
modularized code, 106
package store, 109, 110
reflection on setup, 112
source repository, 106
testing, 107–109
training pipeline trigger, 112
user data collection, 111
Continuous model delivery, 87, 95
    automated model building, 97
    automated training pipeline, 99
    data analysis, 96
    deploy pipeline, 99
    feature store, 95, 96
    model registry, 100
    model services, 100
    modularized code, 98
    performance, 101
    reflection on setup, 102–104
    training pipeline trigger, 101
Convolutional neural network
    (CNN), 6
Credit card data set
    kaggle website page, 10
    loading data set, 11–15
    normal and fraudulent, 16–18
    packages, 11

Printed in the United States
By Bookmasters